Drones in Society

The integration of drones into society has attracted unprecedented attention throughout the world. The change, for aviation, has been described as being equally as big as the arrival of the jet engine. This book examines the issues that surround this change, for our society and the legal frameworks that preserve our way of life. *Drones in Society* takes the uninitiated on a journey to understand the history of drones, the present day and the potential future in order to demystify the media hype.

Written in an accessible style, *Drones in Society* will appeal to a broad range of interested readerships, among them students, safety regulators, government employees, airspace regulators, insurance brokers and underwriters, risk managers, lawyers, privacy groups and the Remotely Piloted Aircraft System (RPAS) industry generally. In a world first, this book is a light and interesting read; being both relatable and memorable while discussing complex matters of privacy, international law and the challenges ahead for us all.

Ron Bartsch has over 30 years' experience in the aviation industry in a variety of senior operational, safety and regulatory roles. Ron is President of the Asia Pacific RPAS Consortium and is a former Director of the Australian Association of Unmanned Systems (AAUS) and recently completed his thesis on the legal aspects of RPAS operations with the University of Sydney. As former Head of Safety and Regulatory Compliance at Qantas Airways Limited and a senior manager with the Australian Civil Aviation Safety Authority, Ron has operational, safety and regulatory experience across all sectors of the aviation industry. Ron is also a presiding member of the Administrative Appeals Tribunal as an aviation specialist.

James Coyne is a former Civil Aviation Safety Authority senior manager and the former Co-chair of the International Civil Aviation Organisation Unmanned Aircraft Systems Study Group. He is currently the Technical Director of UAS International. Jim holds a BEng with Honours and an MSc in Aero-systems Engineering. He also serves on the Board of the Australian Association of Unmanned Systems.

Katherine Gray has over 20 years' experience as a regulator, lead auditor and project manager. She has a strong aviation background and a keen interest in human factors. Katherine is the Senior Associate of UAS International, the primary independent unmanned aircraft systems (UAS) accreditor in the Asia Pacific region. She holds a BEng in Aerospace Engineering and an MBA.

Drones in Society

Exploring the strange new world of
unmanned aircraft

**Ron Bartsch, James Coyne and
Katherine Gray**

Routledge
Taylor & Francis Group

LONDON AND NEW YORK

First published 2017 by Routledge

2 Park Square, Milton Park, Abingdon, Oxfordshire OX14 4RN
711 Third Avenue, New York, NY 10017

Routledge is an imprint of the Taylor & Francis Group, an informa business

First issued in paperback 2018

British Library Cataloguing in Publication Data
A catalogue record for this book is available from the British Library

Library of Congress Cataloging in Publication Data
Names: Bartsch, Ronald I. C., 1954– author. | Coyne, James, 1952–
author. | Gray, Katherine, 1972– author.
Title: Drones in society : exploring the strange new world of unmanned
aircraft / Ron Bartsch, James Coyne, Katherine Gray.
Description: Abingdon, Oxon; New York, NY: Routledge, 2017. |
Includes bibliographical references and index.
Identifiers: LCCN 2016029565| ISBN 9781138221574 (hardback) |
ISBN 9781315409658 (ebook)
Subjects: LCSH: Drone aircraft.
Classification: LCC TL718 .B37 2017 | DDC 629.133/39—dc23
LC record available at https://lccn.loc.gov/2016029565

ISBN: 978-1-138-22157-4 (hbk)
ISBN: 978-1-138-36291-8 (pbk)

Typeset in Bembo
by Florence Production Ltd, Stoodleigh, Devon, UK

Contents

Abbreviations

AC	Advisory Circulars
ACAS	Airborne Collision Avoidance System
ADS–B	Automatic Dependent Surveillance-Broadcast
AGL	Above Ground Level
AMA	Academy of Model Aeronautics
ANC	Air Navigation Commission
AOC	Air Operator's Certificate
AROC	Aeronautical Radio Operator's Certificate
ASTM	American Society for Testing and Materials
AT&T	Aircraft Transport and Travel
ATM	Air Traffic Management
ATS	Air Traffic Services
AUVSI	Association for Unmanned Vehicle Systems International
AVIC	Aviation Industry Corporation of China
BRIC	Brazil, Russia, India and China
BVLOS	Beyond Visual Line of Sight
C2	Command and Control
CAA	Civil Aviation Authority
CAB	Civil Aeronautics Board
CAP	Civil Aviation Publication
CASA	Civil Aviation Safety Authority
CASR	Civil Aviation Safety Regulations
CofA	Certificate of Airworthiness
COA	Certificate of Waiver or Authorization
COTS	Commercial-Off-The-Shelf
CPDLC	Controller–Pilot–Data-Link–Communications
CS	Certification Specification
CS–LUAS	Certification Specification–Light Unmanned Aircraft System
CS–LURS	Certification Specification–Light Unmanned Rotorcraft System
CTA	Consumer Technology Association
DAA	Detect and Avoid
DGCA	Directorate General of Civil Aviation

DOT	Department of Transport
EASA	European Aviation Safety Agency
EU	European Union
EUROCAE	European Organization for Civil Aviation Equipment
FAA	Federal Aviation Administration
FAI	Fédération Aéronautique Internationale
FPV	First Person View
GAF	(Australian) Government Aircraft Factory
GPS	Global Positioning System
GPWS	Ground Proximity Warning System
ICAN	International Commission for Air Navigation
ICAO	International Civil Aviation Organization
ICC	International Criminal Court
ICJ	International Court of Justice
IFR	Instrument Flight Rules
JAA	Joint Airworthiness Authorities
JAR	Joint Aviation Requirements
JARUS	Joint Authorities for Rulemaking on Unmanned Systems
KE	Kinetic Energy
LIDAR	Light Detection and Ranging
MAA	Model Aeroplane Association
MAAA	Model Aeronautical Association of Australia
MFC	Model Flying Club
MIT	Massachusetts Institute of Technology
MOA	Memorandum of Agreement
MoU	Memorandum of Understanding
NAA	National Aviation Authorities
NAS	National Airspace System
NATO	North Atlantic Treaty Organization
NHLS	National Health Laboratory Service
NPRM	Notice of Proposed Rulemaking
PICAO	Provisional International Civil Aviation Organization
RPA	Remotely Piloted Aircraft
RPAS	Remotely Piloted Aircraft System
RPASM	Remotely Piloted Aircraft System Manual
RPASP	Remotely Piloted Aircraft Systems Panel
RPL	Remote Pilot License
RPS	Remote Pilot Station
SARPs	Standards and Recommended Practices
SES	State Emergency Services
SOR	Summary of Response
SVS	Synthetic Vision System
TC	Type Certificate
TCDS	Type Certificate Data Sheet
TSO	Technical Standard Order

UAS	Unmanned Aircraft System
UASi	UAS International
UASSG	Unmanned Aircraft Systems Study Group
UAV	Unmanned Aerial Vehicle
UCAV	Unmanned Combat Air Vehicle
UK	United Kingdom
UN	United Nations
UOC	Unmanned Aircraft System Operator's Certificate
US	United States
USAF	US Air Force
VFR	Visual Flight Rules
VHF	Very High Frequency
VLOS	Visual Line Of Sight

1 The rise of the drone
Introduction

> If you change the way you look at things, the things you look at change.
> Wayne Dyer

The introduction of drones into society has been described as being as significant and revolutionary as the advent of the jet engine. Some commentators go even further. It has been suggested that the drone—or unmanned aircraft as they are more commonly referred to in civilian life—is arguably the greatest aviation innovation since the Wright brothers' Flyer back in 1903. And what is even more amazing about drones is the fact that any aircraft flying today has the ability to be flown without a pilot onboard. So are pilotless aircraft to become the horseless carriage of the twenty-first century?

Compared to the challenges that accompany the introduction of any radical new technology, such as the cell phone or the internet, the integration of unmanned aircraft into society presents even greater challenges. Managing change in the context of a highly technological and rapidly changing society has, even since the advent of the traditional piloted aircraft, been the most challenging role of governments and aviation regulators. However, with the rapid emergence of drones into society, unique issues arise that challenge existing assumptions and regulatory models.

Apart from what drones have done and continue to do for the military, they have already proven themselves as capable of conducting surveillance, patrolling borders, agricultural spraying, searching for missing persons, photography and inspection of emergency situations like bushfires, floods and cyclones. And yes, of course, delivering pizzas. Amazon has estimated that about 85 percent of what they sell online weighs less than two kilograms (5 pounds), and that the drones currently available would be more than capable of delivering your order to your doorstep.

Across all of these applications the issues of safety, privacy and ethics often become topics of debate even before the technology is deployed. While regulators and their regulations may limit the use of drones for commercial and civil purposes from a safety perspective, the major societal issues have been more related to the privacy implications of their usage. Moreover, the unmanned aircraft that seems to be causing most concern in respect to invasion of privacy are those that are being used for recreational purposes, with the remote "pilot" controllers having little or no aviation experience or exposure.

The global economic potential of the civil application of drones is astounding. A recent study estimated the worldwide market for unmanned aircraft systems (UAS) at over US$150 billion in 2016. Advocates are keen to point out the many and diverse range of ways that they are going to make our lives better. Gretchen West, Senior Advisor at Hogan Lovells US LLP and former Executive Vice President for the US Association for Unmanned Vehicle Systems International (AUVSI) recently stated, drone technology is a tremendous tool to increase efficiencies across almost all industries. But their potential extends far beyond the domain of traditional piloted aircraft operations.

Many people seem convinced, and with good reason, that unmanned aircraft are about to take over our lives. The reality is a little more complicated. Although drone activity around the world is increasing at an exponential rate in the civilian sector, when it comes to commercial usage, many countries are struggling with how to regulate them. Consequently many states have applied strict limitations and in some instances total bans. But as was so clearly demonstrated with alcohol prohibition in the USA during the 1920s, if there is overwhelming demand for a commodity or service within society, laws and regulations become ineffective and can indeed have the opposite affect to their original purpose. Drones seem to fit squarely into this category.

For governments, corporations and individuals alike, it is suggested that a paradigm shift may be required to effectively integrate drones into society. Scary stuff—especially considering the dizzy pace at which drone technology continues to evolve. But it is not only the rapid rate of development of UAS technology that requires a recalibration of approach as to how we are to control this new invention. Rather it is the unique characteristics, capabilities and diversity of UAS applications.

If recent drone events are evidence of what the future may hold then we are only at the beginning of the consciousness of human imagination as to what drones may ultimately be capable of. And where the human imagination goes, lawyers will soon follow. Unmanned aircraft give rise to a myriad of legal issues, and all of them, like the technology, are in evolution.

Unmanned aircraft technology is evolving faster than our ability to understand how, legally and ethically, to control it. However, if we as a society cannot effectively control drones then, apart from the harm, hazards and hindrance they may unleash upon us, any positive contribution or advantages they may afford—and there are many—cannot and will not be fully realized.

For good and for bad

The technology for drones appears enticingly cheaper than piloted aircraft and therefore provides a direct cost saving. These savings and benefits also extend to the environment through reduced aircraft emissions. But the integration of drones into society might be more restricted and restrictive than first envisaged. Fundamentally, their potential cannot be fully realized unless and until they are able to share the same domestic airspace as the rest of commercial aviation.

In the past, as with any new and internationally prolific invention, societies and their governments—through their rule-making capacities—have harnessed such technologies for the betterment of society in general. Acknowledging that the sinister, evil, perverted and criminal element will always remain, so long as regulations, and our law enforcement agencies, can keep pace with the technology, the good will always outweigh the bad. Credit card fraud and other cyber crime will persist and so the regulators must remain persistent. The main problem confronting governments in respect to drones is that they "invaded" well before any effective regulatory regime was established. It's certainly the case of the drone having literally slipped under our radar of consciousness.

Unmanned aircraft nevertheless remain divisively different from previous flying machines and other inventions. Seeing drones in their operating domain is one thing but the experience of operating this radically new and deeply strange technology is something else. According to some operators the encounter is somewhat surreal.

> A drone isn't just a tool; when you use it you see and act through it— you inhabit it. It expands the reach of your body and senses in much the same way that the Internet expands your mind. The Net extends our virtual presence; drones extend our physical presence.
>
> (Grossman, 2013)

Unlike previous technological developments drones are more *agile, accessible, affordable, adaptable* and more capable of *anonymity*. These "five A" attributes of unmanned aircraft are confronting society and their governments throughout the world. Legislators, public interest groups and certain sectors of society are resistant to the integration of this strange, new technology. They fear, and with good reason, intrusions into their everyday life and privacy violations.

Contrasting drones with traditional manned aircraft becomes even more apparent when one considers their agility. Piloted aircraft are very much constrained by what may be considered as their natural limits. Drones have no such constraints, as they have no limitations. Just when we think we have imposed boundaries to harness the invention—technology unleashes the beast. Drones resist all possession. They are capable of a swarming, persistent low-level presence, but above all anonymously. Whereas at times traditional piloted aircraft may be considered to be a necessary inconvenience, many consider drones as an unnecessary evil. Pervasive and uncontrolled or perhaps even worse—*incapable* of being controlled.

As the rampant pace of drone innovation continues, together with its attributes of agility, accessibility, affordability, adaptability and anonymity, with it comes the potential for widespread abuse by users. The rate of drone developments and the commensurate potential for invasion of privacy, together with the threat to national, corporate and personal security, has understandably generated considerable public debate.

In the past the courts have been slow to react when addressing new technologies—credit cards, automatic tellers and the internet—all of which left society and the general public somewhat vulnerable through abuse and misuse of the new technology. It has been mainly left to governments and regulatory authorities to impose limits and restraints upon the technology and in some instances prohibit their use entirely. The rapid emergence of unmanned aircraft has been no exception. Governments' *modus operandi* in such situations seems to be: "if we can't control them—ban them." But alas, the drones have escaped and there's no putting this airborne genie back in the bottle.

The dawn of the age of the drones and the potential they hold, for good and for bad, provides a new challenge, a challenge to which the law needs to catch up. The genie cannot be constrained but the genie nevertheless needs to be controlled. At present the drone dilemma is very much a case of "unmanned and uncontrolled."

Now we know the nature of the problem that faces society and governments throughout the world, let's look at the phenomenal rate of proliferation of this strange new technology.

The drones have landed and multiplied

It is now trite to say that drones represent a game-changing development for the aviation industry. While it is estimated that only 2 percent of current expenditure on UAS in the world encompasses civilian UAS, that percentage is expected to increase exponentially over the next decade. But still there are more unmanned aircraft in the world today than their manned counterpart. In the US, AUVSI estimates that upon integration of UAS into the National Airspace System (NAS), and by 2025, more than 100,000 jobs will have been directly created in the US alone and provide an economic impact of US\$82 billion.

UAS are poised to become part of everyday air services operations perhaps within the next few years. Needless to say there are significant challenges that need to be addressed in order to seamlessly introduce UAS into civilian airspace. In those countries that have drone regulations, most operations are restricted in their operations to segregated airspace—clear of commercial air transportation. At present only about one-third of the 191 countries that are bound under international aviation conventions have enacted drone regulations.

Another challenging aspect of this new type of aircraft is that, since their inception, UAS have become smaller, more sophisticated and increasingly less expensive. Their application is as varied as their design. As was described earlier in this chapter, the rapid pace of UAS technological development can be directly related to what is described as its "five 'A' attributes" of agility, accessibility, affordability, adaptability and anonymity.

In many counties, and in particular in the US, the most apparent and immediate application of UAS has been in conducting surveillance. This fact is of little surprise considering this was the military application for which drones were originally developed. For example, the Global Hawk drone was initially developed for the US Air Force toward the end of the 1990s and was first used shortly after the terrorist attacks the World Trade Center on September 11, 2001. NASA's new Hurricane Hunters are the drones that tracked Bin Laden. Many UAS are fitted with high-resolution cameras and imaging technologies. The research and development arm of the UAS industry is likewise growing exponentially.

Currently there are hundreds of types and designs of UAS of both fixed wing and rotary variants. Drones range in size from insect-like micro-drones to large commercial aircraft. "Lethal" and "unobtrusive" micro-drones are currently being developed by the US Air Force to mimic the behavior of bugs. April 2013 saw the first commercial air transport aircraft to be operated as a Remotely Piloted Aircraft System (RPAS). With two persons onboard, the Jetstream 31-type aircraft flew 500 miles in the non-segregated UK airspace sharing the sky with other commercial airliners. On that day a new chapter in aviation history was written. This revolutionary flight reflects the rapidly expanding use of RPAS for civilian use and simultaneously raises several complex legal issues.

It is this diversity of aircraft design and consequent diversity of application that is raising considerable legal issues. As one commentator (Farber, 2014) suggested: "The defence and aerospace industries are propelling UAS into our lives faster than the courts and lawmakers can prepare for their ubiquitous and powerful presence."

According to one leading international drone consultancy firm, UASi, in 2016 in the USA alone, nearly 100 companies manufactured approximately 250 different drones. Worldwide expenditure on drones was in excess of US$8 billion. In a major report released in May 2016 by PwC, the world's largest professional services firm, the estimated market value of drone-powered solutions is in excess of US$127 billion (Mazur and Wisniewski, 2016). In the USA, it is estimated that by 2020 some 30,000 UAS will be occupying the US national airspace. Further it is predicted that in the same year US$11.4 billion will be spent per annum on UAS sales with a cumulative total of US$89 billion over the next decade. The UAS sector is expected to create 70,000 new jobs in the first three years of integration into the NAS and over 100,000 jobs by 2025.

The increase in sales of drones has been nothing short of spectacular. In 2014 Australian UAS manufacturer MultiWiiCopter told the Australian Parliament's "Eyes in the sky" inquiry that its local client base included more than 5,000 customers, and consumer UAS vendor Parrot claimed that it has sold 500,000 units globally. Although the market for civil use currently comprises less than 2 percent of the worldwide market for unmanned aircraft, that could change over the next few years as technology advances and as legislation and regulations allow broader use of unmanned aircraft in unsegregated civilian airspace.

There is absolutely no doubt that there are significant operational and societal benefits that UAS can provide in terms of cost savings, mission diversity and potential reduction in environmental emissions in comparison to piloted aircraft operations. One area in which UAS can provide significant advantages, as compared with manned aircraft, is in the area of reconnaissance and surveillance, and this has been the major usage of drones in the US experience. UAS have the ability to be cost-effective while simultaneously improving the efficiency of law enforcement and reducing the risk to law enforcement officials.

The availability of UAS provides domestic law enforcement agencies with the opportunity to operate more efficiently by obtaining otherwise unavailable surveillance information that could potentially lead to greater and more accurate arrests. The use of facial recognition technology is highly effective from a drone vantage point—but equally controversial. In many countries, including the USA and Australia, both the federal and the State police forces have utilized UAS in hostage scenarios with positive outcomes. UAS operations also allow for advanced national security with increased border patrol and coastal surveillance capability, as well as the opportunity to conduct more sophisticated and effective emergency surveillance and search and rescue missions. Trials of drones for coastal surveillance and border security in the US and many other countries have been undertaken for the past few years.

In a disaster, drones can be quickly deployed to assess affected areas and improve situational awareness, aiding first responders in determining the most effective allocation of resources and delivery of life-saving supplies. Firefighting services are using UAS equipped with thermal imaging equipment to provide advanced intelligence in respect of the detection and spread of forest and bush fires.

Of the various categories of work potentially open to UAS, other than surveillance applications, an economic study commissioned by AUVSI suggests that precision agriculture and public safety present the most viable commercial markets, at least in the US.

What has become a popular and apt descriptor for UAS applications is that they can be used for "dull, dirty and dangerous" missions where it would be perilous to assign a human pilot for such missions. In his paper, Marshall (2007) describes the history of unmanned or remotely piloted aircraft (RPA), the technological challenges facing operators of such systems, and the unique legal and regulatory issues that have arisen because of the rapid evolution of this new, but not-so-new, sector of aviation. In his own words Marshall describes the phrase as follows:

> "Dull, dirty and dangerous" is a common description of the potential uses and utility of unmanned aircraft, operated as military surveillance and communications platforms, hardened weapons delivery systems, observation and interdiction assets for national security and border protection, and any number of civilian or nonmilitary applications. Scientists use them for intercepting and measuring atmospheric phenomena such as hurricanes,

sampling the air quality over disaster areas, and flying through volcanic eruptions where manned aircraft would risk loss of aircraft and human life, a few of the many current and envisioned aviation missions. Any current activity in which airborne assets are deployed in a "dull" (long endurance, high altitude, fatigue-inducing), "dirty" (volcanic plumes, chemical spills) or "dangerous" (high risk, low altitude such as firefighting) environment may potentially be conducted in a safer, less expensive, and more efficient manner with unmanned aircraft.

As a concluding comment to this section, it appears that the potential scope of UAS application is almost limitless. Even the use to date of drones in certain applications far exceeds the capability or efficiencies of traditional piloted aircraft operations. The operational costs of UAS alone are significantly lower than their manned counterparts and some have significantly greater endurance— particularly those that are solar powered or solar assisted. One highly acclaimed UAS commentator (Michaelides-Mateou and Erotokritou, 2014) lists a range of civilian applications to include border protection in support of immigration control, law enforcement and homeland security, agricultural use, aerial photography, search and rescue, disaster management (for example the Fukushima power plant explosion in 2011) and leisure activities.

The future use and application of drones seems to be limited only by our imagination. But what effect has this rampaging technology had on society and how have our established societal controls and values coped? In the following section we will explore the impact of integrating this strange new invention into contemporary society. This presents unprecedented challenges at various levels: into modern society generally; into our legal system; into the existing civil aviation regulatory framework; and into unsegregated civilian airspace. Indeed this invention we call drones essentially represents a fusion of science, society and the law.

Relationship between the law, technology and aviation

It was the culmination and convergence of three different technologies that led to the explosion of drone innovation and their proliferation. First there have been dramatic improvements in the efficiencies of drone power plants, whether it be lightweight electric motors or micro piston-driven or jet engine technology. In respect of electrically powered devices, these have been accompanied by commensurate improvements in battery technology.

The second important area of technological development has been in the advancement and availability of Global Positioning Systems (GPS). Their incorporation into everyday life and delivery through our mobile phones and smart devices has been staggering. To be able to determine, with pinpoint accuracy, our position on this planet would have amazed and bewildered our navigational-centric forefathers. Drones avail themselves of this GPS technology and, with their three-dimensional agility, are able to freely locate and perform their mission—whatever that may be.

The final area of scientific advancement has been the remarkable improvements in lightweight cameras with their high-resolution images and versatile operations. Kodak knows only too well—in hindsight—the rate of progress of digital camera technology. Kodak engineer Steven Sasson developed the first digital camera in 1975 with a resolution capability of 0.01 megapixels. Today, many small drones are fitted with 36 megapixel capacity cameras—a staggering 3,600-fold increase in resolution capacity compared to Kodak's first digital camera.

Apart from these three areas of rapid technological advancement, the widespread usage of smart devices, especially mobile phones, provide for seamless and no-cost controlling device platforms through the simple downloading of an app. All these areas of technological development have soared while their costs have been consistently reducing. Attempts in the past to develop useful terrestrial robot-like technology have had limited success. It seems that the clumsy terrestrial robot has finally found its feet—albeit in the sky.

It is the freedom and agility by which aviation operations can readily transcend previously restrictive geographic and political boundaries that truly differentiates flying from all other modes of transport. With respect to drone operations, this "freedom and agility" transcends to an entirely new and unprecedented level. To harness this freedom for the betterment of all, aviation regulation over all forms of aviation—whether of the manned or unmanned variety—provides the requisite authority, responsibility and sanctions. The regulation of aerial activities is as fundamental and rudimentary to the aviation industry as civil order is to modern society. In no other field of human endeavor or branch of law does there exist such a vital yet symbiotic relationship.

It is no revelation that aviation and regulation are intrinsically linked. In fact it is generally recognized that aviation is the most strictly and extensively regulated industry. The above view is highlighted in the following extract:

> The aviation industry is what it is today not in spite of, but rather because of, the law that regulates it.
>
> (Bartsch, 2012)

But with the UAS generation, as with previous eras of rapid advancement of aircraft design, performance and capability, the law is lagging behind science and technology. The ongoing development and use of unmanned aircraft highlights Ravich's (2009) observation:

> Law lags science; it does not lead it. UAS exemplify the modern information age, an era of computer automation, the Internet, high-definition imagery, and "smart" technology. More can be done virtually and by remote control today than at any time in history, and the corresponding actual and potential savings of personnel and resources are tangible. In aviation parlance, UAS are the leading edge of contemporary aeronautical science and engineering and a product of a century of manned flight experience.

However, UAS operations have outpaced the law in that they are not sufficiently supported by a dedicated and enforceable regime of rules, regulations, and standards respecting their integration into the national airspace.

The capacities of UAS are increasing while their affordability and thus presence do likewise. The law is undoubtedly having difficulty keeping up but how long can a regulatory regime based on exceptions and limitations handle the volume of demand for this strange new aircraft type? It is debatable as to whether one regulatory regime can provide both standards-based rules to account for safety concerns and also provide adequate privacy safeguards. Then there are the other areas of the law that involve security, corporate espionage and terrorism that need to be addressed. Regulations of any kind must be sufficiently flexible to provide beneficial applications of the technology that are not currently contemplated. Drones, like other robots, will continue to challenge lawyers and legal policy makers in the semi-foreseeable future. Legislators have to address the issue of integration and get out in front of it.

Those aspects of integration that are at present creating the greatest challenges derive from the differences between UAS and any previous type or variety of aircraft design and technology. Therefore, from a legal and regulatory perspective, the real integration difficulties stem from the inherent differences and diversities of this new form of aircraft combined with the fact, obviously, that UAS are piloted remotely and in some limited instances, fully autonomous. And perhaps the most sobering aspect of all is that the rate of advancement of UAS technology is showing absolutely no signs of abating and indeed is continuing to increase.

Sharing the sky

As previously stated unmanned aircraft or drones—as they were traditionally referred—were conceived and developed within a military milieu. Unmanned Aerial Vehicles (UAVs) are, as with many previous aviation innovations—including the jet engine—a child of war. It is only a matter of time before we begin to experience their full potential within the civilian and commercial aviation sector. Unlike military drone activities, civilian operations, particularly commercial usage, cannot rely on specially designated and restricted military airspace. Therefore, one of the key issues for operating UAS for civilian purposes will be their integration into non-segregated common airspace.

Civilian UAVs are intended to operate in a different environment than military UAVs, namely in the common, non-segregated civilian airspace, together with all other air traffic. What is acceptable in military operations and in segregated airspace does not apply to civilian applications. At the moment, UAVs lack formal airworthiness certification by civilian aviation authorities. There are no airworthiness standards and acceptable means of compliance for those technical features of UAV technology, which go beyond traditional manned aircraft. The main obstacle for civilian UAVs to fly in non-segregated airspace is safety.

(Kaiser, 2011)

As civilian usage of UAS becomes more and more common, and their commercial and operational superiority more demonstrable, the pressure imposed upon governments and regulators for access to unsegregated and ultimately unrestricted civilian airspace will intensify. Of course there will be public outcry from certain sectors of society and individuals but their cost-benefit superiority will most likely prevail. Such is the nature of democratic societies. Ultimately UAS will be flying with the same degree of freedom as manned aircraft, in non-segregated airspace. Therefore a solid legal framework on the international, regional (for example, the European Union (EU) and the European Aviation Safety Agency (EASA)) and national level laying down all technical, safety and operational requirements will need to be implemented.

Until now, regulatory authorities have been focusing most of their attention on the issue of developing and imposing limitations on drone usage and have made comparatively few proposals with regard to the creation of an entirely new regulatory framework for their use in common (unsegregated) airspace. Perhaps a better regulatory approach, and one that has been advocated by the International Civil Aviation Organization (ICAO), is to first consider all drones as aircraft and then identify the differences—the "delta"—in their operations to that of their piloted counterparts. In adopting such an approach, the existing, mature and "safety-proven" regulatory structure can remain and incorporate drones into that regime, as and when the identified "deltas" are safely addressed.

The diversity of application of UAS usage that differs from that of manned aircraft is what places the adequacy of existing law and regulatory framework under so much pressure. We therefore need to consider the scope of UAS activities and applications (possibly by way of a gap analysis) to ensure that any changes to the regulatory regime will be sufficient to accommodate and harness UAS technology now and well into the future. An ABC journalist, Mark Corcoran, who is an expert on UAS technology, stated: "I think that the problem is that the technology is now progressing at such a rate that regulators and legislators risk being buffeted in the slipstream."

The enormity of the task of integrating UAS into unsegregated civil airspace cannot be overstated, however the legal issues associated with UAS activities are not restricted to safety and technical regulation. While the advantages to society in general of increased usage of UAS is undeniable, because of the uniqueness of this new type of aircraft technology, its increased civilian usage also raises a number of important legal, social and ethical issues.

Most UAS activities, where regulators have permitted operations, are currently constrained to segregated airspace such as low-level airspace (below 400ft above ground level or AGL), test sites, designated danger areas or within temporary restricted areas. On some occasions, UAS operations are permitted in an extremely limited environment outside segregated airspace. To exploit fully the unique operational capabilities of current and future UAS and thus realize the potential commercial benefits of UAS, there is a necessity to be able to access all classes of airspace and operate across national and international borders and airspace boundaries. Such operations must achieve an acceptable (equivalent) level of safety but regulation should not become so inflexible or burdensome that commercial benefits are either frustrated or denied.

Many regulatory authorities throughout the world have adopted the above approach as they transition to a more "safety systems" and less prescriptive approach to safety regulation. The emphasis is placed more on the operator to demonstrate to the regulator, usually through the development of an exposition, that they have established robust safety systems and procedures that ensure they can carry out their operations safely. The following extract from the Australian Civil Aviation Safety Authority (CASA) clearly describes the commercial advantages of such an approach:

> Apart from the safety benefits that will be derived from this new regulatory regime clearly articulating the safety outcomes that are required, the new regulations will allow greater operational flexibility for airlines in achieving these safety outcomes. This legislative flexibility will provide significant opportunities for organisations in terms of integrating safety into their business planning processes.

As a result of the vast diversity and uniqueness of UAS operations, a highly prescriptive aviation regulatory framework, that is common throughout the world, is simply not practical or feasible to encompass the scope of drone applications. If the legislation is structured in such a way that it prescribes the desired safety outcomes, the operator has the flexibility to structure their procedures so that they are both safe and commercially sustainable.

The viability of the commercial market for UAS, especially in the civil market, is heavily dependent on unfettered access to the same airspace as manned civilian operations to enable sustainable commercial operations. While it is essential that UAS demonstrate an equivalent level of safety compared to manned operations, the current regulatory framework has evolved around the concept of an onboard pilot or pilots. There is a need to develop UAS solutions that assure an equivalent level of safety for UAS operations, which in turn will require adaptation or transition of the current regulatory framework to allow for the concept of the remote pilot without compromising the safety of other airspace users.

One of the major issues facing UAS operations is the demonstration of equivalence (in particular for detect and avoid systems) in the context of an evolving air traffic management (ATM) environment. It is very important to understand that the current ATM environment is not static. Achieving equivalence with manned operations is not a fixed target as there are many significant changes proposed that aim to improve operational efficiency and performance or enhance safety. On the whole, proposed changes to the ATM environment could be seen as advantageous to UAS operations as more and more functions within the environment are automated, thus there is a significant opportunity for the UAS industry to influence the shape of the future ATM environment to support wider UAS operations.

Summing up

It is impossible to accurately predict the true form of technological advances, or the impact they will have on aviation, society and our legal system. It is equally difficult to envisage the impact or risks they may impose upon our way of life. Contemporary society has proven to be remarkably adept at integrating and normalizing technological developments, especially once any moral anxiety relating to their introduction subsides. On the other hand, the negative impacts of some technological advancements have only become apparent subsequent to their introduction and integration into society, which makes them much harder to regulate and control.

The challenge for governments throughout the world is to ensure that the risks associated with the introduction of this new technology are managed and balanced in such a manner that permits society to benefit as a whole. The technical standards that UAS will need to meet to be integrated into non-segregated airspace will be considered in some detail in later chapters. It is expected that new technology will resolve many of the issues associated with UAS, including maintaining separation, establishing airworthiness, maintaining contact with air traffic control, loss of link or sight of UAS procedures, and ensuring the security of the link and base station. There epitomizes a touch of irony in that technology created the "drone" issue but may also prove to be the solution.

As these technical challenges are overcome, authorities will need to be aware of the implications for other stakeholders with aviation responsibilities. For example, it has been suggested that air traffic controllers are best placed to provide separation advice for UAS operators, which arguably effects a de facto transfer of responsibility for separation from the pilot of the UAS to air traffic control. Issues relating to situational awareness and other non-technical aspects of drone operations require special consideration. The human factors element of drone operations is considered in some detail in Chapter 8.

Authorities will also need to ensure that new technologies have demon-strated efficacy and that evidentiary requirements are met. For example, the UAS command and control (C2) link must be secure and able to detect and monitor when deliberate interference has been attempted. Equally, larger drones may require onboard tamper-proof instruments to record the actions of the remote pilot. In imposing these additional requirements, authorities will have to balance the need for safety against the cost—as per usual cost-benefit analysis.

When is a drone not a drone? Terminology and definitions

In terms of researching and discussing any new technology it is most important to have an accepted definition of the subject matter. Public familiarity with unmanned aircraft, popularly known as drones, comes largely from their use in military operations abroad. Unmanned aircraft or UAVs are aircraft operated without the possibility of direct human intervention from within or on the aircraft. The simplicity of this definition belies, however, the complexity of classification of unmanned aircraft for regulatory purposes, an issue explored in more detail in the following section.

There is considerable ambivalence about the use of the term "drone" within this sector of the aviation industry mostly because of its negative military connotations with targeted killing. Lay commentators and media personnel have been corrected and even reprimanded by the unmanned aircraft sector for failing to use terms like unmanned aircraft system (UAS), UAVs or RPAS. As this is a book about drones it is important that we should discuss the various terms, their origin and how best to describe them in their civilian capacity.

In 2005 ICAO decided upon the use of the term "UAV" and defined it as "a pilotless aircraft—which is flown without a pilot-in-command on-board and is either remotely and fully controlled from another place . . . or programmed and fully autonomous." A couple of years later in 2007 ICAO dispensed with the use of "UAV" and decided upon "UAS" as the preferred term and defined it as "an aircraft and its associated elements which are operated with no pilot on board."

Another two years later in 2009, ICAO, in adopting a recommendation of their UAS Study Group, first introduced the term "remotely piloted aircraft" or RPA. This adaptation was based on the conclusion that only unmanned aircraft that are *remotely* piloted could be integrated alongside manned aircraft into non-segregated airspace (that is airspace used by commercial aviation) and at aerodromes. From that time on ICAO decided to narrow its focus from *all* UAS to those that are remotely piloted.

The use of the descriptor "remote" or "remotely" as an adjective allows for standard ICAO definitions to continue to be used. Therefore the definition of a pilot remains unchanged with a *remote* pilot being "a person charged by the operator with duties essential to the operation of a remotely piloted aircraft and who manipulates the flight controls, as appropriate, during flight time." Similarly a *remote* pilot-in-command is "the remote pilot designated by the operator as being in command and charged with the safe conduct of a flight." A *remote* co-pilot is a "licensed remote pilot serving in any remote piloting capacity other than as remote pilot-in-command but excluding a remote pilot serving in remote piloting capacity for the sole purpose of receiving flight instruction." And finally a *remote* flight crewmember is a "licensed remote crew member charged with duties essential to the operation of a remotely piloted aircraft system during a flight duty period."

In adopting such taxonomy this provides an efficient way in which the *differences* in operating RPA can be readily identified. In other words it becomes an exercise in determining the "delta," that is, that which differs from operating UAS as compared to traditional aircraft operations with the pilot (and other crewmembers) onboard.

The term most commonly used in the USA and by the Federal Aviation Administration (FAA), and increasingly used by the general community and the media, is "unmanned aircraft system" or UAS. The FAA Modernization and Reform Act of 2012 distinguishes between the aircraft and the associated systems by defining an unmanned aircraft as "an aircraft that is operated without the possibility of direct human intervention from within or on the aircraft." On the other hand "UAS" refers to the airframe as well as the associated communication links and control station and is defined by the FAA as:

> An unmanned aircraft and associated elements (including communication links and the components that control the unmanned aircraft) that is required for the pilot in command to operate safely and efficiently in the national airspace system.

As was stated previously, if the broader benefits that unmanned aircraft operations can provide are to be fully realized, they must be fully integrated into unsegregated airspace both domestically and internationally. Therefore, ultimately, UAS terminology must align throughout the world if there is to be international harmonization. For that reason, and with ICAO having finalized their *Remotely Piloted Aircraft Systems Manual* (RPASM) in 2015, the following definitions from that manual will be used:

- *Remotely piloted aircraft (RPA)*—an unmanned aircraft which is piloted from a remote pilot station; and
- *Remotely piloted aircraft system (RPAS)*—a RPA, its associated remote pilot station(s), the required C2 links and any other components as specified in the type design.

The RPASM further defines a "RPAS operator certificate" as "a certificate authorizing an operator to carry out specified RPAS operations." Although the term "UAS" is listed in the manual's glossary of acronyms the term is not defined in the RPASM. As will be discussed in the following section, because the term UAS is broader than RPAS and encompasses all types of unmanned aircraft operating systems, the previous extracted (US Congress) definition of UAS will be adopted throughout this publication.

The practical reality of RPA or unmanned aircraft operating in the same airspace as manned civil aircraft requires these aircraft to have ability to act and respond in real-time and comply with instrument flight rules (IFR) and visual flight rules (VFR) as per piloted aircraft. Currently, there is no recognized technology solution that could make these aircraft capable of meeting regulatory requirements to see (detect) and avoid other aircraft, and for the C2 integrity to provide the RPA with the equivalent level of safety of an aircraft with a pilot onboard.

Depending on the situation, unmanned aircraft may be required to "recognize" aerodrome signs and markings, identify and avoid terrain, identify and avoid severe weather, provide visual separation from other aircraft and avoid collisions. Article 3 of the Chicago Convention also requires civil aircraft to be able to comply with a State's request to land or deviate to a designated airport. One of the greatest challenges of full UAS integration into non-segregated airspace is technology enabling the remote pilot to identify, in real-time, the physical layout of the aerodrome so as to maneuver the aircraft safely.

It is these technical aspects of unmanned operations, together with the legal issues associated with determining and apportioning liability, that pose the greatest challenges generally but in particular in respect of fully automated operating systems.

Fully autonomous systems

A major technical, legal and regulatory challenge of the "pilotless" element is full automation of flight. The safety of fully autonomous flight necessarily becomes a legal question in terms of the degree of flight control that the related computers are permitted to have and where legal liability might reside. Automation of flight has been progressing at astonishing rates in the past few decades and fully autonomous UAS may still be some way off but will eventually be developed.

One of the most visible effects of automation in the operation of aircraft is the progressive reduction of the number of flight crew in commercial air transport operations. In some instances the reduction has been from six flight crewmembers down to two and potentially, at least technologically, zero. By way of example, post-World War II commercial airline operations typically used to operate with six technical crewmembers consisting of a pilot-in-command (captain or commander), a co-pilot (first officer), second officer, flight navigator, flight engineer and radio operator. By way of comparison the minimum technical (flight) crew component for the world's largest commercial aircraft—the Airbus A380—is just two pilots.

Full automation will be the pinnacle of the development of unmanned flight and is strongly spurred by the technical advancement of communication and information technologies and their miniaturization. Remote control and fully automated operations can complement each other in a way that autonomous flight can provide redundancy if a control link fails. Even though full auton-omous flight operations maybe feasible in the future, both ICAO and the US Congress consider that there must be human responsibility for, and with authority over, the flight of the unmanned aircraft. Accordingly, mostly based on legal argument, it must be ensured that the responsible remote pilot-in-command can override the autonomous flight mode at any time, when necessary.

In terms of ICAO terminology, a RPA or RPAS cannot, by definition, include fully autonomous systems. If this were the case then a consideration of RPAS in not including fully autonomous systems would impose considerable limitations of what is a rapidly developing area of UAS research and technology. For this reason, and to allow consideration of the entire scope of potential unmanned operations, autonomous systems will be included in this publication and therefore both terms—RPAS and UAS—will be used as applicable.

The ICAO RPASM defines an "autonomous aircraft" as "an unmanned aircraft that does not allow pilot intervention in the management of the flight." The ICAO RPASM goes on to define an "autonomous operation" as "an operation during which a remotely-piloted aircraft is operating without pilot intervention in the management of the flight"—which significantly does not preclude the existence of a "remote" pilot.

In summary the two families of terms differ to the extent that RPA/RPAS excludes fully autonomous flight, whereas UA/UAS may include or exclude them. As the US Congress requires a pilot-in-command, fully autonomous UAS are excluded from consideration of the FAA—at least at this point in time. The presence and requirement of a remote pilot has led ICAO to define this UAS-type as a RPAS for the reasons described above.

Although the term "drone" continues to be used as a generic term it is more commonly used to describe the military application of unmanned air-craft operations. With an increasing number of safety regulators throughout the world categorizing small RPAS (less than 2kg/5lb) and requiring less stringent regulation—these are increasingly being referred to as "drones"—and also by ICAO. Significantly, however, the word "drone" does not appear throughout the entire ICAO RPASM. The ICAO, as stated in the RPASM, now considers the term UAV as being "obsolete" and do not define the term in the Manual.

Elsewhere throughout the world other aviation regulators, for example the EASA and the Australian CASA, in following ICAO approach, use the terminology "remotely piloted aircraft" in respect of the civilian usage of un-manned aircraft. The unmanned aircraft or RPA together with the data (C2) link and ground control units (remote pilot stations) are, as with ICAO, collectively and respectively referred to as UAS or RPAS.

A concluding comment

Since its very beginning, aviation was a pioneering and creative industry but, at least in its early years, at the expense of many accidents and the cost of many lives. Today, aviation is based on mature and robust safety systems and cultures. Traditionally it takes years, if not decades, to develop a new aircraft type. Innovations are integrated slowly, through the application of strict regulatory controls in accordance with internationally accepted safety and technical standards. The result is a very good safety record.

It is highly attractive to merge the innovative power and competitive pricing of information technologies with aviation. Unmanned aircraft are a product of such a merger. But society continues to struggle to accommodate and integrate unmanned aircraft as governments and their legal systems are overwhelmed by the rate of technological advancement. It is contended that if these aircraft are to be integrated into common, unsegregated civil airspace, then under current laws relating to aircraft certification standards, and in accordance with international conventions, they necessarily must be subject to the same, or at least equivalent, technical and safety standards.

As previously stated, any aircraft flying today is capable of being flown by a remotely located pilot. Therefore if the current level of safety of our air transport system is to be maintained there must be at least an equivalent level of safety for all unmanned operations if they are to share the same airspace.

Public awareness of drones and the political and societal understandings of them are thought to be paramount if there is to be support and acceptance of the technology and its future development. The introduction of UAS into domestic airspace raises never-before faced safety and integration issues. Of the many challenges facing governments throughout the world, and also for UAS manufacturers and operators seeking commercial opportunities in civilian markets, perhaps the most significant is the lack of airspace regulation that covers all existing and contemplated unmanned systems and operations.

Access to the common airspace is the key to everything regarding unleashing the full potential of drones and the benefit they can bestow upon society. To achieve this goal, the focus of new regulations must be to maintain the safety levels the aviation industry has achieved during the past century of piloted operations.

The regulation of drones continues to develop and poses several legal and regulatory challenges. Different jurisdictions have approached such challenges in particular ways depending on national objectives and priorities. Examples from the US, Europe and Australia show that the safety of air navigation is of paramount concern and has naturally attracted close attention. The current prohibition that remains in many countries on the commercial usage of UAS may have inhibited research and development in this area and ultimately, for these nations, impacted upon the timeliness of their integration into society.

It is clear that the regulatory challenge for all countries is the assurance of at least an equivalent level of safety for all drones in common national airspace. There is a crying need for the development of harmonized international rules relating to UAS operations, but the difficulties and challenges remain. This chapter has attempted to provide an overview of the emergence of drone operations into the civil aviation industry and create an awareness of some of the main issues—technical, legal and ethical—that this transition has created in the broader community. UAS are unique and therefore may require unique solutions to the problems and challenges of this strange yet exciting new technology.

References

Bartsch, Ron *International Aviation Law* Ashgate, Farnham, UK, 2012.

Clothier, R "Pilotless Aircraft: The Horseless Carriage of the 21st Century" (2008) 11(8) *Journal of Risk Research* 999–1023.

Farber, Hillary B "Eyes in the Sky: Constitutional and Regulatory Approaches to Domestic Drone Deployment' (2014) 64(1) *Syracuse Law Review* 1.

Grossman, Lev "Game of Drones" *Time*, February 11, 2013.

House of Representatives Standing Committee on Social Policy and Legal Affairs "Eyes in the Sky: Inquiry into Drones and the Regulation of Air Safety" The Parliament of the Commonwealth of Australia, Canberra, July 2014.

Kaiser, Stefan A "UAVs and Their Legislation into Non-segregated Airspace" (2011) 36 *Air and Space Law* 162.

Marshall, Douglas M "Dull, Dirty, and Dangerous: The FAA's Regulatory Authority Over Unmanned Aircraft Operations" (2007) 5 *Issues Aviation Law and Policy* 10.

Mazur, Michal and Wisniewski, Adam "Clarity from Above" PwC Global Report on the Commercial Applications of Drone Technology, PwC Poland, May 2016, p. 4.

Michaelides-Mateou, Sofia and Erotokritou, Chrystel "Flying into the Future with UAVs: The Jetstream 31 Flight" (2014) 39 *Air & Space Law* 111.

O'Sullivan, Emma "New Technologies Present New Challenges for Law Makers" *About the House*, August 2014, Commonwealth of Australia, Canberra, Australia, p. 45.

Ravich, Timothy M "The Integration Of Unmanned Aerial Vehicles Into The National Airspace" (2009) 85(597) *North Dakota Law Review*.

2   From battlefield to backyard

Development of unmanned aircraft

The secret of getting ahead is getting started.

Mark Twain

Many of the conflicts around the globe today in war zones such as Afghanistan, Iraq and Syria against Al-Qaeda and Islamic State, involve the use of UAS such as the General Atomics MQ-1 Predator (reconnaissance, combat), General Atomics MQ-9 Reaper (reconnaissance, air attack), the Northrop Grumman RQ-4 Global Hawk (reconnaissance), Boeing Insitu, ScanEagle (reconnaissance) and the Israel Aerospace Industries Heron (reconnaissance).

The military view of the introduction of unmanned aircraft as a life-saving development for their own personnel as they have put an end to any immediate danger to their pilots and other flight crew by removing them from the onboard cockpits and thus the active war zone and have them now operate the unmanned aircraft from a remote location. Similarly, politicians are keen to embrace UAS, as the up-front purchase costs are generally lower than manned aircraft (compare the 2013 costs of a MQ-9 Reaper at approximately US$17 million with the approximate cost of an F-35 Lightning II at US$70 million). Unmanned aircraft are also stealthier, have the ability to fly at higher altitudes and are generally less expensive to operate than manned aircraft. UAS maintenance costs are less as the equipment is generally commercial-off-the shelf (COTS) and they can fly without the stringent safety requirements required in manned aviation. Further, operating costs are less expensive. Compare the Mesa County landfill project in the US. It costs the County US$10,000 for a contractor to fly a manned aircraft to survey the site, whereas the job can be done using an unmanned aircraft for $300.

One of the main advantages of unmanned aircraft over manned aircraft from a military perspective is their ability to stay in the air for longer periods of time. An average sortie for a modern fighter aircraft is of the order of approximately two to three hours. Compare this with the MQ-9 Reaper, which has an endurance of between 25 to 30 hours. The difference depends on whether the Reaper is carrying a full weapon load or simply conducting surveillance missions. General Atomics, the manufacturer of the MQ-9 Reaper, has assessed that it could almost double the endurance by adding fuel pods and longer wings. Imagine the strategic advantages of an unmanned aircraft fitted with missiles, or a bank of surveillance equipment that could remain up in the air over an area of conflict for nearly two days.

The beginning

So, where and how did it all start? While it is hard to believe, the earliest recorded use of an unmanned aircraft occurred in 1849. It occurred in a military conflict at a time when the Austrian Empire was being threatened by a number of revolutionary movements, which took place between March 1848 and November 1849. One such revolutionary State was the Republic of San Marco. The Republic had been established following a revolt in Venice against Austrian rule in March 1848. The Austrians laid siege to the Italian city of Venice, which led to starvation and outbreaks of cholera in the city. During this siege, the Austrians attacked Venice and launched the first air raids in history. This was first accomplished on July 15, 1849 from unmanned free balloons that were loaded with explosives and then flown over Venice. Subsequent bombardments continued and the Republic of San Marco surrendered to Austria on August 22, 1849. The Austrians had been working on this system for months (Holman, 2009).

The "Presse" of Vienna reported:

> Venice is to be bombarded by balloons, as the lagunes (sic) prevent the approaching of artillery. Five balloons, each twenty-three feet in diameter, are in construction at Treviso. In a favorable wind the balloons will be launched and directed as near to Venice as possible, and on their being brought to vertical positions over the town, they will be fired by electro magnetism by means of a long isolated copper wire with a large galvanic battery placed on a building. The bomb falls perpendicularly, and explodes on reaching the ground.

Further, the following account appeared in the *British Morning Chronicle* a week after the surrender:

> The Soldaten Freund publishes a letter from the artillery officer Uchatius, who first proposed to subdue Venice by ballooning. From this it appears that the operations were suspended for want of a proper vessel exclusively adapted for this mode of warfare, as it became evident, after a few experiments had been made, that, as the wind blows nine times out of ten from the sea, the balloon inflation must be conducted on board ship; and this was the case on July the 15th, the occasion alluded to in a former letter, when two balloons armed with shrapnels ascended from the deck of the Volcano war steamer, and attained a distance of 3,500 fathoms in the direction of Venice; and exactly at the moment calculated upon, i.e.,

at the expiration of twenty-three minutes, the explosion took place. The captain of the English brig Frolic, and other persons then at Venice, testify to the extreme terror and the morale effect produced on the inhabitants. A stop was put to further exhibitions of this kind by the necessity of the Vulcan going into docks to undergo repairs, which the writer regrets the more, as the currents of wind were for a long time favourable to his schemes. One thing is established beyond all doubt, viz., that bombs and other projectiles can be thrown from balloons at a distance of 5,000 fathoms, always provided the wind be favourable.

(Holman, 2009)

In 1863, just two years after the start of the American Civil War (1861–1865), a New York City-based inventor, Charles Perley, developed the "Perley Aerial Bomber," which was a hot-air balloon that carried a container full of explosives attached to a crude timing mechanism. At a pre-determined period, the timer would tip the basket, causing the explosives to fall out and the fuse to ignite. Unfortunately, the bomber was inaccurate and dangerous, and it is unclear whether it was ever used successfully.

In relation to these unmanned free balloons, according to ICAO, for an unmanned aircraft to be regarded as such, it must be piloted, i.e. controlled, albeit from a remote location. Consequently, an unmanned free balloon does not meet the definition of a UAS as it cannot be managed on a real-time basis during flight and is thus deemed to be uncontrolled.

Another candidate that may have been considered as an early UAS occurred in 1887 when British meteorologist, Douglas Archibald, attached a camera to a kite that he used to take the first aerial photographs (Archibald, 1897). The kite, which is traditionally tethered to the ground or to the person controlling it, had an additional string that was attached to the shutter-release of the camera. Since kites are tethered, so that they don't fly away uncontrollably, they do not constitute a UAS in accordance with ICAO definition.

To pilot an unmanned aircraft from a remote location requires it to be operated and controlled by a radio signal using a remote control device. Radio control requires the following parts:

- Transmitter—This can be as simple as a hand-held device or as complicated as a full mission control room, depending on the complexity of the UAS. The transmitter is used to control the unmanned aircraft and generates an electric current that is fed into an oscillator circuit that produces an electro-magnetic wave, known as the radio signal. The transmitter then sends the radio signal via an antenna to the receiver.
- Receiver—This is a device inside the unmanned aircraft that has an antenna matched to the transmitting antenna and receives radio signals from the transmitter that it then converts to electric power to activate servo motors inside the unmanned aircraft as commanded by the transmitter.

- Servo motors—Servo motors can turn wheels or gears, adjust flaps, operate propellers, etc. Servo motors are different from normal electric motors, which rotate an arbitrary number of times according to how long they receive an electric current. Servo motors are much more controllable and have built-in electronic feedback mechanisms that enable them to rotate by reasonably precise amounts.
- Power source—The power source is typically provided by batteries, which can be normal batteries or a rechargeable battery pack.

When we take the cockpit and thus the pilot out of the aircraft, it is clear that radio is to be an integral part of the UAS in order to sustain controlled flight, and many engineers and scientists dedicated their efforts to experimenting with radio in the very early days. Nikola Tesla (July 10, 1856–January 7, 1943), a Serbian American physicist, inventor, engineer and futurist, created the first basic design for a radio. In the northern spring of 1893, Tesla addressed the Franklin Institute in Philadelphia and the National Electric Association at St. Louis, where he described in detail the principles of radio broadcasting. He also made the first public demonstration ever of radio communication. Next to Tesla, another scientist most deserving of credit for pioneering radio was the British physicist, Sir Oliver Joseph Lodge (June 12, 1851–August 22, 1940), for in 1894 he demonstrated the possibility of transmitting telegraph signals wirelessly by Hertzian waves over a distance of 150 yards (Cheney, 2001).

Two years later in 1896, Guglielmo Marconi (April 25, 1874–July 20, 1937) arrived in London with a wireless set identical to Lodge's. The set also had a ground connection and antenna exactly the same as Tesla had described in his widely published lectures of 1893 (Martin, 1894). Marconi's apparatus, which transmitted radio signals for about 2.4km, received a patent for radio in 1896. Marconi has been called the "father of radio," as a result of his claim for his radio invention that was awarded the first patent. Marconi gained further acclaim as he also shared the 1909 Nobel Prize in Physics with Karl Ferdinand Braun "in recognition of their contributions to the development of wireless telegraphy." Tesla tried on many occasions to prove that he was actually the inventor of radio, but did not have the backing or finances that Marconi had to be able to substantiate his claim. However, on June 21, 1943, the Supreme Court of the United States handed down a decision on Marconi's radio patents deeming them invalid and restoring some of the prior patents of Sir Oliver Joseph Lodge, John Stone Stone (sic), an American mathematician, physicist and inventor, and Nikola Tesla. Although Tesla was a man ahead of his time, to this day people still have very little idea about Tesla's work with radio, or indeed many of his other inventions. The US Supreme Court decision is dated April 23, 1943, three months after Tesla's death. Although, in acknowledging Marconi's work, it would be fair to say that Marconi was the "father of long-distance radio communication" (Bondyopadhyay, 1995).

Another little known fact about Tesla is that he predicted the development of the mobile phone in 1909. In an interview with the *New York Times*, published in *Popular Mechanics*, Tesla stated:

> It will soon be possible, for instance, for a business man in New York to dictate instructions and have them appear in type in London or elsewhere. He will be able to call from his desk and talk with any telephone subscriber in the world. It will only be necessary to carry an inexpensive instrument no bigger than a watch, which will enable its bearer to hear anywhere on sea or land for distances of thousands of miles.
>
> (1909)

Of course, the work of Tesla, Marconi, Braun et al., all built on the earlier works of James Clerk Maxwell, Hans Christian Ørsted, André-Marie Ampère, Joseph Henry, Michael Faraday and Heinrich Rudolf Hertz. In fact, the first systematic and unequivocal transmission of electromagnetic waves was performed by Hertz and described in papers published in 1887 and 1890, many years before either Tesla's or Marconi's work. Hertz was the first to conclusively prove the existence of electromagnetic waves, which had been theorized by Maxwell's electromagnetic theory of light, which he presented to the Royal Society of London on December 8, 1864 and later published in his *A Dynamical Theory of the Electromagnetic Field* (1865). Hertz proved the theory by engineering instruments to transmit and receive radio pulses using experimental procedures that ruled out all other known wireless phenomena. The unit of frequency—cycle per second—was named the "hertz" in his honor (International Electrotechnical Commission, 1930).

Still on Tesla's work, on November 8, 1898, he patented the first radio-controlled robot. The device was a boat that he called a "teleautomaton." He demonstrated this boat during an electrical exhibition at Madison Square Garden in 1898 (O'Neill, 1944). With the success of his radio-controlled boat, Tesla tried to sell his idea to the US military as a type of radio-controlled torpedo, but they showed little interest. Remote radio control remained no more than a novelty until World War I, and then all of a sudden, a number of countries saw the light and began to use it for military purposes.

The first unmanned aircraft

Following closely in Tesla's footsteps was an engineer, research physicist and inventor by the name of Archibald Montgomery Low (October 17, 1888–September 13, 1956), who developed some brilliant radio control techniques. Low earned the title "the father of radio guidance systems" because of his ground-breaking work on radio remote control capabilities that are still used to this day for missile guidance systems, guided rockets, remote controlled planes and torpedoes.

Low served in the Royal Flying Corps, the precursor of the RAF, where he was tasked with finding a way to remotely control an aircraft so it could be used as a guided missile. This project was called "Aerial Target" and the prototype was the "Ruston Proctor Aerial Target," which is generally regarded to be the first "pilotless" aircraft (Taylor, 1977). It had its first flight on January 1, 1916, only 13 years after the first successful controlled manned flight by the Wright brothers on December 17, 1903.

Low also developed the world's first wireless rocket guidance system in 1918. Unfortunately, like so many other ground-breaking inventions, Low's inventions were under-appreciated and the British government took very little notice at the time. However, never fear, that was not the end of it as we shall see later.

Development of unmanned aircraft

The development of unmanned aircraft has largely kept pace with the development of conventionally piloted aircraft. So, it is not surprising that once the airplane had been invented and had proven itself capable of controlled flight, the technological developments that would allow them to be flown as unmanned variants for military purposes were not far behind.

Soon after the development of the Ruston Proctor Aerial Target, the Hewitt-Sperry "Automatic Airplane," otherwise known as the "flying bomb," made its first flight on September 12, 1916. These unmanned aircraft were intended to be used as "aerial torpedoes," which can be considered to be an earlier version of today's cruise missiles. Control was achieved using gyroscopes developed by Elmer Sperry of the Sperry Gyroscope Company.

A demonstration of the capabilities of the automatic airplane was provided to the Aviation Section of the US Army Signal Corps in November 1917, which resulted in the commissioning of a project to build an "aerial torpedo." The result was an experimental, unmanned aerial torpedo called the Kettering Bug, which was named after its designer Charles Franklin Kettering, of the Dayton-Wright Company (Cornelisse, 2002). The Kettering Bug was capable of striking ground targets up to 120km away and could travel at speeds of 80kph. The first flight of the Kettering Bug took place on October 2, 1918. While the Kettering Bug was a successful project that proved the capabilities of the technology, it was never fully developed in time to be deployed in the war, which ended on November 11, 1918. Apart from the "Bug," Kettering made many valuable contributions to the automotive and electronics industries with his inventions, which included the electrical starter motor, leaded gasoline and Freon refrigerant.

Following World War I, many conventionally piloted aircraft were converted into radio-controlled "pilotless" aircraft for use as targets or guided missiles. These included the converted US Army Standard E-1 (1917), and the RAE Larynx cruise missile (1925). Many new technologies were being experimented with at that time, as seen by the huge variety of UAS developed in the 1930s and 1940s. In 1931 the British developed the Fairey "Queen," a remotely controlled target drone, which was a derivative of the Fairey IIIF floatplane.

Probably one of the most successful unmanned aircraft was the DH.82 Queen Bee. Although urban legend has it that the Queen Bee was a converted Tiger Moth, it actually wasn't. The aircraft used the basic wooden fuselage of a DH.60GIII Moth Major with DH.82A Tiger Moth wings, but the whole aircraft was highly modified to permit catapult operations and was fitted with floats or a wheeled undercarriage depending on whether it was launched from land or water. Later in the war a handful of genuine Tiger Moths were modified to the Queen Bee standard but probably only as an experiment. The aircraft was flown under wireless control for the training of anti-aircraft gunners on ships and on land. While the aircraft was classified as a target, it was not strictly so, as the intention was to shoot "near" the aircraft but not directly at or to hit it. Most of the aircraft written off were as a result of flying out of range of wireless control and were lost or damaged from heavy landings back on the open sea, rather than from being shot down (McKay, 2014).

The Queen Bee project was highly classified and trials had been underway for almost two years before the Air Ministry invited the press to a demonstration at Farnborough in 1935 (McKay, 2014). The Queen Bee was a game changer in the unmanned aircraft world as it was the first multi-use unmanned aircraft that was designed as an aerial target for anti-aircraft gunners training purposes rather than as a single-use guided weapon. The Queen Bee could fly at altitudes of 17,000ft at speeds over 160kph, and was capable of traveling distances of over 400km. It is generally accepted that the term "drone" was adopted in homage to the Queen Bee.

By the outbreak of World War II, unmanned aircraft were given designations based on their design and function: "A" denoted "Attack drones," "PQ" designated "full-sized" target drones, and "OQ" denoted a "small-scale" target drone.

The first large-scale production, purpose-built unmanned aircraft was developed by Reginald Denny, whose full-time career was as a stage, film and television actor. In his leisure time, he was also an avid aviator and a pioneer in unmanned aircraft. Denny was born in the UK in 1891 and served with the British Royal Flying Corps during World War I. Following the war, he migrated to the USA to seek his fortunes in Hollywood as an actor. In the 1930s, Denny became interested in radio-controlled model airplanes and in 1934 opened a model aircraft shop on Hollywood Boulevard, known as the Reginald Denny Hobby Shop. The shop later became the site of the Radioplane Company.

Denny believed that one of the best uses for a radio-controlled unmanned aircraft was as a training aid for anti-aircraft gunners, and in 1935 he demonstrated a prototype target unmanned aircraft, known as the RP-1, to the US Army. Denny bought an airplane design from Walter Righter in 1938 and after successfully building the aircraft he began marketing it to hobbyists as the "Dennyplane." He also demonstrated it to the Army, using the designator

"RP-2." This military prototype underwent several iterations and in turn became the "RP-3" and "RP-4." In 1940, Denny won a US Army contract for his radio-controlled RP-4, which became known as the OQ-2." Denny's company manufactured nearly 15,000 OQ-2s for the US Army during World War II. The Northrop Corporation purchased Denny's company in 1952. Northrop went on to merge with Grumman in 1994 to become the Northrop Grumman Corporation, which, *inter alia*, developed the Global Hawk, which is in operation today and is also the largest unmanned aircraft in production.

The Radioplane Company was doing some very good work for the war effort and employed many technical staff to assemble the OQ-2. A young lady by the name of Mrs Norma Jeane Dougherty was a technician at the company, where she was working as part of the war effort. A photographer named David Conover was assigned by an Army publicist, Captain Ronald Reagan, to take some photographs of Mrs Dougherty and later persuaded her to work as a model, which was the beginning of her career. She later changed her name to Marilyn Monroe. Incidentally, Captain Reagan became the 40th President of the USA in 1981.

Another little known inventor who contributed to the future of unmanned aircraft, and who was also a Hollywood star, was Hedy Lamarr, an Austrian–American actress who was known as "the most beautiful woman in the world" during her career. Lamarr, with the assistance of George Anthiel, invented frequency hopping and spread spectrum, a concept for having wireless radio signals switch quickly among many different frequency channels so that the signal cannot be jammed. Anthiel and Lamarr, under her real name of Hedy Kiesler Markey, received a "Secret Communications System" patent in 1942. Unfortunately, the US Navy ignored the invention, probably because they didn't understand it. Fortunately, the patent surfaced again in the mid-1950s and the Navy suddenly realized its significance. Frequency hopping and spread spectrum techniques are used extensively today in both civilian and military technologies and it is used effectively in the unmanned aircraft industry. Today, even basic, inexpensive unmanned aircraft have the ability to switch between various frequencies, which has the added bonus of rendering signal jammers ineffective.

Early radio-controlled unmanned aircraft could only operate within visual line of sight (VLOS) of the remote pilot. Edward M. Sorensen invented a radio-controlled unmanned aircraft that could fly beyond visual line of sight (BVLOS) and the remote pilot could determine the aircraft altitude, attitude and flight dynamics from the ground station. The application for a patent for his radio remote control system was filed on May 16, 1940 and was issued on October 8, 1946. Unfortunately, Sorensen died before the patent was issued.

Getting back to the inventions of Archibald Montgomery Low, while the British government took very little notice of his work in those early years, by 1944, the Germans were well aware of his inventions and made good use of his rocket guidance system as one of the foundations for their Vergeltungswaffe -1 (V-1) flying bombs, which were launched against Britain during the latter part of World War II. A total of 9,521 V-1s were fired at the British with over 100 a day being launched at the peak of activity. A further 2,448 V-1s were fired against Belgium until the last launch site was overrun on March 29, 1945. The V-1 could travel a distance of 240km, reach speeds of almost 800kph and drop nearly a tonne of explosives. It is generally accepted that V-1s killed nearly 1,000 civilians and injured more than 35,000 in British cities alone.

After World War II, France produced an aerial target, which was essentially a copy of the V-1, albeit slightly smaller with twin tail surfaces. It was known as the CT-10 and could be launched from an aircraft or from the ground, using a rocket booster. Some CT-10s were sold to the UK and the US. The USA also built its own version of the V-1, which was known as the Jet Bomb (JB)-2 Loon. The JB-2 was developed for use against Japan in Operation Downfall, which was planned to start in October 1945, but never saw combat as the Japanese surrendered on August 15, 1945.

During this post-war period, Australia was conducting its own research into unmanned aerial targets. The Australian Government Aircraft Factory (GAF), located at Fishermans Bend in Melbourne, Victoria, built two manned proto-types to test the aerodynamics, engine performance and radio control systems parameters as a proof of concept. The prototype was known as the "Pika," which is an Australian Aboriginal word meaning "flyer." Development of the Pika began in 1948 with the first flight occurring in 1950. As a result of these trials, Britain became interested in the work being undertaken and signed a bilateral agreement with Australia in relation to guided missile testing. The British provided the missiles to be used in the tests while Australia provided the test facilities, which included the Woomera Test Range in South Australia. As a result of this agreement and following successful trials of the Pika, Australia was awarded a contract to develop an unmanned target aircraft. The unmanned version of the Pika became known as the "Jindivik," which is an Australian Aboriginal word meaning "the hunted one." The first flight of the Jindivik occurred in August 1952.

The Australian GAF produced 502 Jindiviks between 1952 and 1986 for use by the Royal Australian Air Force, the Royal Australian Navy Fleet Air Arm, the Royal Air Force, the Swedish Air Force and the US Navy. The Jindivik was considered so successful that the production line was reopened in 1997 to build another 15 models for Britain.

While many unmanned aircraft were developed from scratch or converted from conventional manned aircraft, their main use was as aerial targets. They did not reach their full potential as something other than a target until the end of the 1950s/early 1960s. At this time they were known as Unmanned Aerial Vehicles (UAVs) and they took to the skies in about 1960 following two significant incidents involving manned spy planes. The first was the shooting down of a U2 over the USSR flown by Captain Francis Gary Powers on May 1, 1960, and the second occurred on July 1, 1960, when a Boeing RB-47 reconnaissance aircraft was shot down near the Soviet border. The program to develop UAVs for long-range reconnaissance missions was known by the code name "Red Wagon." During the war in Vietnam, the US Air Force (USAF) flew approximately 3,435 UAV missions, of which 554 UAVs were estimated to have been lost. In the words of USAF General George S. Brown, Commander, Air Force Systems Command in 1972, "The only reason we need UAVs is that we don't want to needlessly expend the man in the cockpit." Later that same year, General John C. Meyer, Commander in Chief, Strategic Air Command, stated: "we let the drone do the high-risk flying . . . the loss rate is high, but we are willing to risk more of them . . . they save lives!" (UAV and Drone, 2015).

During the Yom Kippur War in 1973, Syrian missile batteries in Lebanon caused heavy damage to Israeli fighter jets. As a result, Israel developed their first modern UAV. The images and radar decoying provided by these UAVs helped Israel to completely neutralize the Syrian air defenses at the start of the 1982 Lebanon War, resulting in zero deaths to pilots.

As unmanned aircraft technologies developed in the 1980s and 1990s, the military became more and more interested in their capabilities. UAVs offered cheaper, more capable fighting machines that could be used without risk to aircrews. Initially, UAVs were used for surveillance activities, but this changed later in the program to include armed variants such as the MQ-1 Predator, which utilized AGM-114 Hellfire air-to-ground missiles. In the military, armed UAVs were known as unmanned combat air vehicles (UCAVs), which later became known as drones.

History of the development of model aircraft

At this time, it is worth spending some time to look at an important subset of unmanned aircraft i.e. model aircraft. The historical development of model aircraft is fascinating. The first model aircraft discovered to date is believed to have been built around 200 BC in Egypt and was uncovered during an excavation of the Saqqara burial grounds, 32km south of Cairo, in 1898 (the same year that Tesla demonstrated his radio-controlled robotic boat in Madison Square Garden). The model is known as the "Saqqara Bird" and measures 150mm long and has a wingspan of 180mm. Obviously, there is a lot of contention as to whether this is actually a model aircraft or simply a child's toy that was made to represent a bird. The controversial notion is more appealing, which suggests that the model could function as a glider as the wings are set at an angle that is similar to that of modern aircraft (Messiha et al., 1991).

Archytas was an ancient Greek philosopher, statesman, strategist, mathematician and astronomer, born around 428 BC and is considered to be the founder of mathematical mechanics. He is also reputed to have designed and built the first self-propelled flying device, which was a bird-shaped steam propelled model, known as "the pigeon" (Reece, 2014). It is recorded to have actually flown approximately 200m before it ran out of steam—pun intended. The flying pigeon has also been called the first robot (Hiskey, 2012).

The first flight by the Wright brothers in 1903, apart from lighting the fire for aircraft designers, manufacturers and pilots, also led to a passion for model airplanes for many aviation enthusiasts. US military engineers used model aircraft to assist in the development of full-sized aircraft that would be used for battle. Aeronautical engineers have been building models to use in wind tunnels to assist with design advancements to aircraft shapes and size, propeller and engine design and to investigate the effects of increased speed for many years.

A model aircraft can be either a static scale model aircraft that sits on someone's shelf or desk, or it can refer to a small unmanned aircraft that actually flies. In either case, the model, whether for static display or for controlled flight, can be a replica of an existing aircraft, an imaginary aircraft or indeed anything that the mind can conjure up. There are models of Superman, witches and bats (the flying type not the type used in cricket or baseball), that have been turned into radio-controlled flyers.

The model aircraft that actually fly come in a variety of types ranging from simple toy gliders made from paper or balsa wood to powered scaled models made from materials such as plastic, Styrofoam, or carbon fibre. The smallest models can weigh as little as a few grams, whereas some can be very large and can weigh several hundred kilograms.

Flying model aircraft are generally controlled by one of the following methods:

- free flight
- control line, or
- radio-controlled.

Free flight model aircraft fly without external control from the ground. The aircraft must be set up before flight so that its control surfaces, center of gravity and weight allow for stable flight. Most free flying models are either unpowered gliders or rubber band powered. This type of model actually pre-dates manned flight. Some variants of these include the Cayley Helicopter Model, designed by Sir George Cayley in 1796, the Cayley Model Glider, 1804 and the Penaud Planophore, designed by Alphonse Penaud in 1871. This latter variant used a twisted rubber band as a power source. Many a child would have played with this type of model aircraft while growing up.

Lawrence Hargrave, an Australian engineer, explorer, astronomer, inventor and aeronautical pioneer invented the Hargrave Box kite in 1893, which helped launch the age of the flying machine. He even managed to attain flight by linking several of his box kites together, creating sufficient lift for him to fly some 4.9 m above the ground at Stanwell Park Beach, New South Wales on November 12, 1894. Professor, Sir Richard Threlfall, in his presidential address to the Royal Society of New South Wales in May 1895, called Hargrave the "inventor of human flight." To honor Hargrave's achievements, a picture of him alongside some of his gliders appeared on the reverse of the Australian $20 banknote from 1966 to 1994. While Hargrave may have achieved human flight some seven years before the Wright brothers, it was not considered controlled flight, so hence he does not hold that honor.

Control line model aircraft are constrained to fly in a circle and are controlled in height by means of inextensible wires attached to a handle held by the person operating the model. Control line models predate manned flight. Early versions had no control and were merely constrained to fly in a circle around a pole. The first person credited with using a recognizably control line model aircraft was Oba St. Clair (April 5, 1912–August 14, 1986) in June 1936, near Gresham, Oregon. The work St. Clair did with model aviation ranks among the most advanced in the twentieth century. His invention of control line powered flying, using a bell-crank, dominated the hobby for almost 40 years. He is considered the true inventor of control line powered flying. A US District Court Judge named Oba St. Clair "The Father of Control Line Flying" (Mackey, 2012).

Control line flying is considered safe when all prescribed safety measures are followed as the airplane is constrained to fly in a circle. A pilot's circle is also provided, and as long as the pilot stays within the pilot's circle and everyone else is outside the outer circle, the model cannot come in contact with persons or property. However, this type of model flying is fairly restricted.

Radio-controlled model aircraft are essentially the same as unmanned aircraft and are controlled in exactly the same manner by an operator on the ground from a remote location using a hand-held radio transmitter. The only difference being is how they are being used. A model aircraft is used for fun and recreation, whereas an unmanned aircraft is used for commercial operations for hire or reward.

The earliest examples of radio guided model aircraft were hydrogen-filled model airships of the late nineteenth century. They were flown as a music hall act using a basic form of spark-emitted radio signal and a method of sequential controls (Boddington, 2004). Miniaturized radio-controlled gear came onto the market in the 1960s, and with it the radio-controlled aeromodeling hobby took off. Scale models for both indoor and outdoor operations were developed.

Flying radio-controlled model aircraft as a hobby grew substantially from the 2000s with reductions in the cost and weight, together with improved performance and capabilities of motors, batteries and electronics. A wide variety of models and styles became available. The most common was the multirotor, which is a rotorcraft having more than two rotors. The more common variants include the quadcopter, hexacopter and octocopter, which refer to multirotors having four, six and eight rotors, respectively. An advantage of multirotor aircraft is the simple rotor mechanics required for flight control. Quadcopters are the most popular type and are relatively inexpensive and very durable due to their mechanical simplicity. Small quadcopters weighing less than a kilogram can be purchased for as little as US$100. Their small size means that they possess low kinetic energy, thus reducing their ability to cause harm to people or damage to property. As quadcopters increase in size and weight, their kinetic energy increases thus increasing their potential for harm.

The hexacopter is similar to the quadcopter but usually larger. They can provide more lifting capacity due to the larger number of motors. It's also possible that if one motor fails, the aircraft can still land safely. The downside is that they tend to be more complex and hence more expensive. The octocopter is an upgraded version of the hexacopter with even more lifting capacity. It is also more reliable because if one or two of the motors fail, it should still be supported by other working motors and able to land safely. However, because octocopters have eight motors, they will draw more current and operators will need to carry extra batteries. They are much more expensive than both the quadcopter and the hexacopter. Multirotors can be fitted with guards that enclose the rotors, which will reduce the potential for harm. However, these guards can reduce the performance of the multirotor.

Due to their low cost, ease of construction and control, quadcopters have become a favorite among amateur model aircraft operators for recreational purposes as well as RPA operators for commercial operations to conduct aerial photography and videos of landscapes and buildings.

Kinetic energy is a useful metric when determining the potential harm from an unmanned aircraft. Kinetic energy is determined from the equation:

Kinetic Energy (KE) = $1/2\ mv^2$
Where m = mass in kilograms
V = impact velocity in meters/second
KE = Kinetic energy in Joules

By using this equation, it is possible to determine a non-lethal kinetic energy and therefore a maximum weight and speed so that, if a collision were to occur with a human, the result is unlikely to be fatal and the probability of it causing a severe injury is also very low. Studies to determine a non-lethal level of kinetic energy have been conducted by Transport Canada, University of Norway, the North Atlantic Treaty Organization (NATO), CASA and the Australian Department of Defence. While there is no definitive conclusion, the general consensus is that being hit by an object with a kinetic energy between 40–60 Joules is not life threatening. Using the above equation, it can thus be determined that an unmanned aircraft weighing 1–2kg traveling at a speed less than 30kph is unlikely to be fatal. The CASA report on the human injury model for small unmanned aircraft impacts conducted in 2013, concluded that "for a 2kg RPA, the highest tolerable velocity for the head impact is below 7.5m/s (30kph)" (Radi, 2013).

Aeromodeling started in Australia in the early 1900s following the invention of the airplane. However, it was not until the 1930s that organized modeling groups appeared. Two rival groups formed in New South Wales, the Model Aeroplane Association (MAA) in 1930 and the Model Flying Club (MFC) in 1931. The two groups merged in 1947 to become the Model Aeronautical Association of Australia (MAAA). It is affiliated to the Fédération Aéronautique Internationale (FAI) through the Australian Sport Aviation Confederation.

While MAAA has no regulatory authority conferred to it by CASA, it is recognized by CASA as a Recreational Aviation Administration Organisation, and has been delegated the administration for flying activities of model aircraft and can issue a number of approvals.

In the USA, the Academy of Model Aeronautics (AMA), founded in 1936, is dedicated to the promotion of model aviation as a recognized sport as well as a recreational activity. AMA works with the FAA concerning model aircraft safety and the operation of model aircraft. It is the largest organization of its kind with a current membership of more than 140,000 members. AMA organizes the annual National Aeromodeling Championships, which is the largest model airplane competition in the world. AMA is also affiliated to the FAI through the National Aeronautic Association (NAA). The NAA was founded in 1905, making it the oldest national aviation club in the USA and one of the oldest in the world. It is a founding member of the FAI.

The link between unmanned aircraft, model aircraft and drones

As defined by ICAO, civil unmanned aircraft, used for commercial purposes, have now become known as RPA, which in turn is just a subset of unmanned aircraft. The RPA is just one component of the total system, which is known as the RPAS. Other components that make up the system include the remote pilot station (i.e. the cockpit), the C2 link and other ancillary equipment such as the launch and recovery systems. This terminology is used to make a definite distinction from model aircraft used for recreational purposes. However, for small unmanned aircraft, there is little if any distinction between an RPA and a model aircraft. As was discussed previously, the difference is purely in the method of operation. A person could own an unmanned aircraft and use it as a model aircraft for recreational purposes on the weekend and then use it as a RPA for commercial purposes during the week where it can be used for hire and reward.

Drones, while they can also be exactly the same as an RPA or a model aircraft, describe a military unmanned aircraft. The major difference is that drones are capable of carrying weapons and by their nature are usually larger and more complex. Notwithstanding, ICAO is now referring to small RPA as drones. The term RPA is reserved for larger unmanned aircraft that are capable of flying internationally.

From battlefield to backyard

The developments in unmanned aircraft from the 1800s used for aerial targets, guided missiles and weapons of modern warfare, or for test models used by the military to help develop better war planes, have now found their way into the backyards and playing fields of children and adults alike to be used for fun and recreation in the 2000s. They are also used for commercial purposes such as aerial photography by real estate agents, movie makers to get better video shots and in the near future, by parcel delivery companies to deliver packages to our homes.

What a journey mankind has taken with these devices, incredible feats of engineering and innovation that initially were not taken seriously by governments or the military or were used to demonstrate "magic" as in Tesla's robotic boat. Whichever way they are looked at, unmanned aircraft, RPA or drones, whether they be used for military purposes, commercial applications or fun and recreation, have always had a place alongside their manned variants and will continue to do so into the future. So, it also true to say that some of these unmanned aircraft have found their way from the battlefield into our backyards.

References

Archibald, Douglas *The Story of the Earth's Atmosphere* George Newnes, 1897.

Boddington, David *Radio-Controlled Model Aircraft* Crowood Press, Marlborough, UK, 2004.

Bondyopadhyay, Prebir K "Guglielmo Marconi – The Father of Long Distance Radio Communication – An Engineer's Tribute" Microwave Conference, September 4, 1995.

Cheney, Margaret *Tesla: Man Out of Time* Simon & Schuster, New York, 2001.

Cornelisse, Diana G *Splendid Vision, Unswerving Purpose: Developing Air Power for the United States Air Force During the First Century of Powered Flight* Wright-Patterson Air Force Base, OH, US Air Force Publications, 2002.

Hiskey, David "The First Robot, Created in 400 BCE, Was a Steam-powered Pigeon" Today I Found Out, November 14, 2012.

Holman, Brett "The First Air Bomb: Venice, 15 July 1849. Airminded – Airpower and British Society, 1908–1941" August 22, 2009.

International Electrotechnical Commission History. http://iec.ch/about/history/ accessed September 13, 2016

Mackey, Charles A Fellow and Associate Historian, AMA. Academy of Model Aviation. "First Documented Successful Control Line Flight" July 20, 2012.

Martin, Thomas C *The Inventions, Researches and Writings of Nikola Tesla, with Special Reference to His Work in Polyphase Currents and High Potential Lighting.* The Electrical Engineer, New York, 1894.

McKay, Stuart *Tiger – The De Havilland Tiger Moth* Crecy Publishing, Manchester, UK, 2014.

Messiha, Khalil et al. "Aeronautics: African Experimental Aeronautics: A 2000-Year Old Model Glider," in *Blacks in Science: Ancient and Modern. Journal of African Civilizations. vol. 5, no. 1–2*, ed. by I. van Sertima (pp. 92–99), Transaction Books, New Brunswick, Canada 1991.

Nikola Tesla interview with the *New York Times* "Wireless of the Future" published in *Popular Mechanics*, October 1909, p. 476.

O'Neill, John J *Prodigal Genius: The Life of Nikola Tesla* Cosimo Classics, New York, 1944.

Radi, Alexander Civil Aviation Safety Authority. "Human Injury Model for Small Unmanned Aircraft Impacts" 2013.

Reece, MR "The Steam-powered Pigeon of Archytas—The Flying Machine of Antiquity" Ancient Origins, 2014.

Taylor, John WR *Jane's Pocket Book of Remotely Piloted Vehicles: Robot Aircraft Today* Collier Books, 1977.

UAV & Drone "Unmanned Aerial Vehicle. History of Unmanned Aerial Vehicles", written by Super User, August 25, 2015, http://uav.altervista.org/history-uav, accessed August 11, 2016.

3   Harnessing the beast

The development of UAS regulations

Laws are like sausages, it is better not to see them being made.

Otto von Bismarck

While the development of unmanned aircraft has largely kept pace with the development of conventionally piloted aircraft, the same cannot be said about the development of the regulatory regime for both types. The development of regulations for manned aircraft followed reasonably closely after the development of the airplane, particularly in countries that were involved with aircraft manufacturing, i.e. the USA and the UK. It is however quite surprising that to date, the USA has only just published its regulation on UAS and the UK does not yet have regulations for unmanned aircraft. The first country to publish an unmanned aircraft regulation was Australia, which occurred in 2002 in the form of Civil Aviation Safety Regulation (CASR) Part 101.

To examine the developments in unmanned aircraft regulations, it is worthwhile looking at the beginnings of the regulatory authorities and the development of manned aviation regulations in both the USA and UK.

USA manned aviation regulations

Remembering that the first successful controlled manned flight by the Wright brothers occurred in 1903, it took another 23 years for the USA to pass the Air Commerce Act which created a brand new Aeronautics Branch within the Department of Commerce. This landmark legislation, approved by President Calvin Coolidge on May 19, 1926, made the Secretary of Commerce responsible for advancing air commerce, issuing and enforcing air traffic rules, licensing pilots, certifying aircraft, establishing airways, and operating and maintaining aids to air navigation.

William Patterson MacCracken, Jr. (September 1888–September 1969) was appointed as the first Assistant Secretary of Commerce for Aeronautics, becoming the first head of the Aeronautics Branch in August 1926 (A brief history of the FAA, n.d.). MacCracken had a keen interest in aviation law and to this day is considered to be a visionary in the world of aviation. He served on the American Bar Association's Committee on Aeronautical Law from 1920 to 1926. Further, from 1922 to 1926, he was a member of the Board of Governors of the National Aeronautical Association. MacCracken came to the attention of President Calvin Coolidge, who selected him to help author the Air Commerce Act. The Secretary of Commerce during the Coolidge years was Herbert Hoover and it was Hoover who appointed MacCracken as his first Assistant Secretary and thus making him the first federal regulator of commercial aviation. The first set of regulations in the USA, known as the Air Commerce Regulations, was written by MacCracken and published in December 1926. Hoover himself went on to become the 31st President of the USA in 1929.

Origins of the FAA

While the Department of Commerce was obviously working to improve aviation safety, a number of high-profile accidents called the department's oversight responsibilities into question. The first occurred on March 31, 1931 when Trans World Airlines (TWA) Flight 599, which was a wooden-winged Fokker F-10 tri-motor airliner, crashed a few miles southwest of Bazaar, Kansas, killing all eight onboard, including popular University of Notre Dame football coach, Knute Rockne. Since the accident is best remembered for causing the death of Rockne, it has become known as the "Rockne Crash" (Friedman and Friedman, 2001). The outcries from the public for greater federal oversight of aviation safety had a profound effect. The accident brought radical changes to aircraft design, operations, airline industry practices and more importantly aviation regulations. The result was a pivotal improvement in airline safety, which resulted in a greater popularity. This single accident radically transformed airline safety worldwide.

In 1934, the Department of Commerce Aeronautics Branch was renamed the Bureau of Air Commerce to reflect the growing importance of aviation to the nation. Unfortunately, the name change did little for advancing safety as the Bureau was ineffectual and hamstrung by poor management. In particular, the reputation of the Bureau came under fire on May 6, 1935, when US Senator Bronson Cutting of New Mexico was killed in the crash of a TWA Douglas DC-2 near Atlanta, Missouri (Glass, 2015). His death was to have nationwide impact in that it would lead Congress to commission an investigation of air traffic safety and the operations of the Bureau of Air Commerce. The Senate appointed Royal S. Copeland, who was the chairman of the Commerce Committee, to head the committee. The preliminary report was issued on June 30, 1936 and gave a scathing account of the Bureau of Air Commerce for providing insufficient funding and maintenance of airway navigation aids.

Following the "Cutting Crash," the Civil Aeronautics Act was passed in 1938. This legislation transferred federal responsibilities for civil aircraft from the Bureau of Air Commerce to a new agency, known as the Civil Aeronautics Authority. The legislation also gave the authority the power to regulate air fares, to determine the routes that air carriers would serve and made it responsible for investigating aircraft accidents.

In 1940, President Franklin D. Roosevelt split the authority into two agencies, the Civil Aeronautics Administration and the Civil Aeronautics Board (CAB). The Civil Aeronautics Administration was responsible for air traffic control, safety programs and airway development, while CAB was entrusted with aviation safety rulemaking, accident investigation and economic regulation of the airlines.

Unfortunately, accidents continued to happen and as a result of a series of mid-air collisions in the 1950s, the most notable being the collision of United Flight 718 (DC-7) with TWA Flight 2 (Super Constellation) over the Grand Canyon, on June 30, 1956, resulting in 128 deaths, the US Congress passed the Federal Aviation Act and it was signed by President Dwight D. Eisenhower on August 23, 1958 (FAA Lessons Learned, 1956). This legislation abolished the Civil Aeronautics Administration and created the FAA. The Act empowered the FAA to oversee and regulate safety in the airline industry and the use of American airspace by both military aircraft and civilian aircraft. Retired Air Force General Elwood Quesada became the first FAA Administrator on November 1, 1958. The FAA began operations on December 31, 1958. In 1966, under President Lyndon B. Johnson, Congress authorized the establishment of the Department of Transportation (DOT), which began full-time operations on April 1, 1967. On that day, the Federal Aviation Agency was renamed the Federal Aviation Administration (FAA), a mere 64 years after the first sustained, powered flight by Orville Wright.

UK manned aviation regulations

The history of the development of aviation regulations in Great Britain can be traced back as far as 1911, when formal control of civil aviation began with the Aerial Navigation Act. Civil aviation law became the responsibility of the Home Office, and the sole concern was the protection of people on the ground. The Act also gave the Board of Trade responsibilities relating to the registration and certification of aircraft and pilots. The Aerial Navigation Act of 1913 transferred control to the Secretary of State for War (Chaplin, 2011).

The World War I years of 1914 to 1918 brought with them a ban on civil aviation except for flights within three miles of a recognized aerodrome. In 1916 an Air Board was formed that was to play a significant role in the post-war control of civil aviation. In the months following the end of World War I on November 11, 1918, a number of regulations were put in place so that civil flying could officially recommence on May 1, 1919 under the supervision of a Department of Civil Aviation, the formation of which had been announced to Parliament on February 12, 1919. In addition to the introduction of the Air Navigation Regulations, a sure fire navigation "belts and braces" system was introduced. This involved painting the names of the railway stations on the roof of the station. Some examples were, Ashford, Hitchin, Redhill and Tonbridge railway stations (Chaplin, 2011).

In his autobiography, Sir Frederick Sykes (1943) notes:

> By dint of superhuman exertions, we managed to get the British Air Navigation Regulations for Civil Flying in the United Kingdom by April 30 (1919). They included rules for the registration of aircraft and the licensing of personnel, the certification of airworthiness for passenger aircraft and their periodic overhaul and examination, registration and nationality marks, log-books, prohibited areas, rules of the air, lights and signals, and customs regulations and other rules for aircraft arriving at and departing from British Airports. Our work was made all the harder by the fact that these regulations were quite novel, and there were no precedents which we could follow.

In 1930 a committee was established under the chairmanship of Air Commodore F.V. Holt to review the certification of aircraft for airworthiness. The issues related to the vast number of different standards that an aircraft had to meet, and the costly and complicated machinery needed to put them into operation. Interestingly, many of the issues made in that submission are the same issues being discussed today about the certification and airworthiness of unmanned aircraft. Further, in those early days the UK was more concerned with the safety of people on the ground, which is also the same issue concerning unmanned aircraft operations today.

In the early years when the rules governing aviation were still being debated in the UK, a lot of attention was devoted to the airworthiness of the aircraft rather than the operations and the operational aspects associated with flying the aircraft. Historically, while operational aspects have caused the majority of accidents, the UK did not follow their USA counterparts in drawing on the experiences obtained from aircraft crashes. In fact, aviation accident reports in the UK in those early days did not make any recommendations on safety. In 1933, a Committee on the control of private flying was established under the Chairmanship of Lord Gorell, who had spent a year from 1921 to 1922 as the Under-Secretary of State for Air, supporting the Secretary of State for Air in his role of managing the Royal Air Force. Despite this, Gorell appears to have learnt very little about aviation safety. The Committee, under his chairmanship took the approach to separate safety and politics. This put the UK out of synchronization with the rest of the world by moving responsibility away from the direct control of politicians. This also meant that the development of aviation regulation in the UK followed a different path from the US.

The Gorell Committee's 1934 report made a total of 18 recommendations, and included an analysis of the causes of accidents to civil aircraft. The report showed that the majority of accidents were due to causes other than airworthiness. One of the conclusions noted:

> Entirely wrong values have been placed upon the relative importance of the pilot, the machine and operational activities, in arriving at the

regulations to be imposed. The failure of the pilot is by far the most potential source of accidents in flying.

However, in spite of the in-depth analysis, the report still concentrated on the airworthiness of aircraft, which doesn't appear to make sense (Gorell, 1934).

In 1936, the UK government created the Air Registration Board, a new body that would examine civil aircraft and issue certificates of airworthiness. The Board comprised four operators, four constructors and four insurers, appointed by the trade interests concerned, together with a representative of the general public and a professional pilot with more than five years' experience as a pilot of civil aircraft, both appointed by the Secretary of State. In addition, four independent members were appointed by the Board itself. The duties of the Board were to advise the minister responsible for civil aviation on matters pertaining to the airworthiness of civil aircraft. This included the approval of modifications to aircraft, the renewal of certificates of airworthiness, daily inspection certificates, etc. The Board established the modern standards of air-worthiness, and during these early years was greatly concerned with the development of Concorde and supersonic flight.

Civil aviation took a back seat during the World War II years of 1939 to 1945. At the end of the war, the responsibilities for civil aviation in the UK were split between three departments; the Ministry of Civil Aviation, the Ministry of Aircraft Production and the Air Registration Board, with each having particular roles and responsibilities. In early 1945, aircraft manufacturers and operators began to receive technical information regarding the investigation of aircraft accidents. Around the same time, the UK also considered that other countries' regulatory authorities should be involved in the investigation of accidents that happened abroad to aircraft on their register.

Finally, as with the US, aviation accidents started to have more direct feedback and helped to shape standards and regulations. An accident involving a London Vickers Viking occurred on September 2, 1958 at Southall, UK, west of London, when it crashed into a row of houses killing all three crew and four people on the ground—a mother and three children. This crash led to the introduction in the UK of the concept of the Air Operator's Certificate (AOC). The AOC ensured that the operator was both competent and properly staffed and equipped, and the regulatory authority employed operationally competent people both to carry out safety checks before the AOC was granted and to supervise its continued validity (UK CAP 789, 2011).

The development of the Concorde had a significant impact on the development of standards and regulations in Europe. Due to the nature of this aircraft being a joint venture between a French aircraft manufacturer, Sud-Aviation (later merged with Nord-Aviation to become Aérospatiale) and the British Aircraft Corporation under an Anglo-French treaty, and the fact that the UK and France did not have a common set of airworthiness standards, led to many challenges. Only 20 aircraft were built between 1965 and 1979 (Towey, 2007). The first flight was on March 2, 1969 and the aircraft was retired on November 23, 2003. Over the years, the UK and France muddled through the design and certification without formalizing any joint regulations. As a result of the ongoing work between the UK and France, in 1970 the Joint Airworthiness Authorities (JAA) was established. Originally, the objectives of the JAA were to produce common certification codes for large airplanes and engines in order to meet the requirements of European industry and international consortia. In 1987, its work was extended to include operations, maintenance, licensing and certification/design standards for all classes of aircraft. The JAA produced a set of Joint Aviation Requirements (JARs), which were developed with a view to minimizing type certification problems on joint ventures, and also to facilitate the export and import of aviation products. The JAA ceased to exist on July 15, 2007 and its functions were taken over by European Aviation Safety Authority (ASA). The JARs later became known as Certification Specifications (CSs) under EASA. The main difference between the JAA and EASA is that the JAA produced harmonized codes without any direct enforcement of law, whereas EASA is a legal regulatory authority. States usually only recognize the regulatory authority of another State. EASA has now changed that, as it is recognized as a regulatory authority in its own right.

Origins of the UK CAA

The UK Committee of Inquiry into Civil Air Transport, chaired by Sir Ronald Edwards, handed down its final report (the Edwards Report) in May 1969. In that report, it recommended that all safety regulations, both operations and airworthiness, should be brought under one authority. As a result of this recommendation, the UK Civil Aviation Act was published in 1971 and the new Civil Aviation Authority (CAA) was established in April 1972 (Edwards, 1969).

Timeline of aviation and regulatory developments

Table 3.1 shows the timeline of the significant events in manned aviation and the subsequent developments in regulations.

Regulation of UAS

As we saw in the previous paragraphs, the regulations for manned aircraft followed reasonably closely after the development of the airplane. The first set of regulations in the US, known as the Air Commerce Regulations, was published in December 1926, 23 years after the first flight. The UK published the Aerial Navigation Act in 1911, although it wasn't until 1936 that the UK Air Registration Board established a set of airworthiness standards (UK Air Registration Board records 1933–1974).

In contrast, the development of regulations for unmanned aircraft has not kept pace with the developments in unmanned aircraft. The Australian CASA

Table 3.1 Timeline of aviation and regulatory developments

Year	Significant events and developments
1903	First successful controlled manned flight by the Wright brothers on December 17, 1903
1911	UK Aerial Navigation Act
1914	Commencement of World War I
1916	UK Air Board established
1919	UK Department of Civil Aviation formed
1921	Australian Civil Aviation (Branch of the Department of Defence) established March 28, 1921
1926	US Air Commerce Regulations, published in December 1926
1931	"Rockne Crash" in the USA
1934	US Bureau of Air Commerce
1935	"Cutting Crash" in the USA
1936	UK Government Air Registration Board
1938	US Civil Aeronautics Administration established
1946	First helicopter crash in the USA
1958	US Federal Aviation Agency
1965	US Recodification CARs/CAMs to FARs
1966	US Federal Aviation Administration (Agency becomes Administration). Placed under the newly created DOT October 1966
1971	UK Civil Aviation Act published
1972	UK Civil Aviation Authority established
2007	European Aviation Safety Agency (EASA) established July 15, 2007

published the first regulation in the world for unmanned aircraft in 2002 in the form of Civil Aviation Safety Regulation (CASR) Part 101. This was 50 years after the first flight of the Jindivik unmanned aircraft in Australia and 86 years after the first flight of the Ruston Proctor unmanned aircraft. A number of other countries have now developed their own regulations for unmanned aircraft, including Sweden, France, Czech Republic, Italy, the Republic of Ireland and New Zealand. The USA published its regulation for small UAS on 28 June 2016, 100 years after the flight of the first unmanned aircraft. The UK has published some very good guidance material, but has not yet published any regulations for unmanned aircraft.

It is important to note that all the regulations published to date relate only to the operational aspects of unmanned aircraft, no regulatory authority has published any standards for certification or manufacture. This gives manufacturers complete freedom to adopt any standard they like, unlike manufacturers of manned aircraft, who must follow a strict regime of standards.

Because regulators are struggling to develop manufacturing and certification standards, other bodies are attempting to fill the void. Some examples of these bodies include the Joint Authorities for Rulemaking on Unmanned Systems (JARUS), which is a worldwide group of regulatory experts from 41 national aviation authorities. The purpose of JARUS is to provide guidance material to facilitate each authority to write their own requirements and to avoid duplicate efforts. To date JARUS has published Certification Specification–Light Unmanned Rotorcraft System (CS–LURS), which is applicable to unmanned rotorcraft under 750kg for operations in VLOS. JARUS is currently working on the development of Certification Specification–Light Unmanned Aircraft Systems (CS–LUAS). The American Society for Testing and Materials (ASTM) International, Committee F38 is developing standards for airworthiness, operations and crew qualifications for UAS.

The rapid growth in the use of unmanned aircraft is a key challenge for regulators. New and innovative ways of using unmanned aircraft are being found daily, ranging from land management and mapping, animal detection and surveillance, crop dusting, telecommunications, cargo transportation, border patrol, law enforcement, providing disaster situational awareness, and supporting search and rescue. There are a number of crucial legal and regulatory issues that must be addressed by national civil aviation regulators and the wider civil aviation international community before unmanned aircraft can be fully integrated into non–segregated airspace.

Within Australia, UAS International (UASi) has developed a standard that encompasses all sectors of the unmanned aircraft industry. The standard has been developed to ensure that all unmanned aircraft are operated safely and in accordance with internationally recognized standards. The standard was developed by safety experts in the field of unmanned aircraft certification and in consultation with the unmanned aircraft industry, former aviation safety regulators and the aviation insurance sector. The UASi standard is the world's first risk–based safety and auditing standard for unmanned aircraft operations. The standard, which is expected to significantly reduce the threat of unmanned aircraft related accidents, was developed by UASi in line with the ICAO RPAS Manual and CASR Part 101.

Although there is a paucity of UAS regulations, aviation regulatory authorities around the world are attempting to integrate unmanned aircraft into civilian airspace using procedural rules and restrictions. Each State has its own way of achieving this by means of guidance material and by utilizing applicable parts of the manned aircraft regulations.

Since we have looked at the development of manned aviation regulations in the USA and UK, it is now worthwhile comparing how these two countries are controlling the operations of unmanned aircraft without actually providing any regulations. We can then compare this approach with those being taken by States that have developed regulations, i.e. Australia, New Zealand and Indonesia.

USA unmanned aviation regulations

For commercial operations, the FAA is taking an incremental approach to the safe integration of UAS into the United States (US) national airspace systems (NAS). As the FAA acquires a better understanding of the operational issues such as training requirements, operational specifications, and technology considerations, it will likely reduce the integration limitations. Until recently, to operate an unmanned aircraft for non-recreational purposes in the US NAS, a user had to obtain a Certificate of Waiver or Authorization (COA), which is an authorization issued by the Air Traffic Organization to a public operator for a specific UAS activity.

The FAA issued Interim Operational Approval Guidance Document 08-01 on 13 March 2008. This document provided guidance on whether an unmanned aircraft was allowed to conduct flight operations in the US NAS and was used to evaluate each application for a COA. While the requirements of each COA was specific to the application, typically the following restrictions applied:

- Flights below 400 ft. above ground level (AGL)
- Daytime operation in Visual Flight Rules (VFR)
- Range limited to Visual Line of Sight (VLOS)
- Greater than 5 miles from an airport

Common uses included law enforcement, firefighting, border patrol, disaster relief, search and rescue, military training and other government operational missions.

FAA policy is based on whether the UAS is to be used for military or civilian purposes or as a model aircraft. An operator who wishes to fly an unmanned aircraft for civil use must obtain an FAA airworthiness certificate the same as for manned aircraft. The FAA is currently only issuing special airworthiness certificates in the experimental category, which cannot be used for compensation or hire. Order 8130.34 outlines the procedures for issuing experimental certificates for UAS. Guidance on the use of model aircraft used for recreational purposes is covered in Advisory Circular 91-57.

By law, any aircraft operation in the US NAS requires a certificated and registered aircraft, a licensed pilot, and operational approval. Section 333 of the FAA Modernization and Reform Act was issued in 2012 and authorizes the Secretary of Transportation to determine whether an airworthiness certificate is required for an unmanned aircraft to operate safely in the US NAS. This authority was leveraged to grant case-by-case authorization for certain UAS to perform commercial operations prior to the finalization of the Small UAS Rule Commercial entities can request relief from the requirements to have a certificate of airworthiness (CoA - not to be confused with a COA) in accordance with the provisions of Section 333. Further, relief from regulations that address general flight rules, pilot certificate requirements, manuals, and maintenance and equipment requirements may also be granted on request. To receive the exemptions, the operator must show that its UAS operations will not adversely affect the safety of persons or property in the air or on the ground, or will provide at least an equivalent level of safety to the rules from which they seek the exemptions. They also need to show why granting the exemption would be in the public interest.

In July 2013, the FAA issued the first ever Type Certificates (TC) in the Restricted Category to Insitu's Scan Eagle X200 and AeroVironment's RQ-20 Puma so that these two small unmanned aircraft could be flown commercially. This was made possible because of the previous military acceptance of the Scan Eagle and Puma UAS designs. This was a significant milestone in the long road of integrating UAS into the US NAS and something for the rest of the world to learn from. In 2015, Insitu conducted flight operations for the US Coast Guard, launching its Scan Eagle from Oliktok Point at the North Slope of Alaska. The demonstration was part of the Coast Guard Research and Development Center's Arctic Technology Evaluation Search and Rescue exercise (SAREX 2015), which was designed to evaluate UAS technologies in remote areas for Search and Rescue (SAR). The Puma's role will be to support emergency response crews for oil spill monitoring and wildlife surveillance over Alaska's Beaufort Sea, within the Arctic Circle.

In December of 2015 the FAA announced that all unmanned aircraft, which includes model aircraft, weighing between 0.55 pounds (250 grams) and 55 pounds (2.5 kg) flown for any purpose must be registered with the FAA. This Interim Rule provides an alternative, streamlined and simple, web-based aircraft registration process for the registration of small unmanned aircraft to facilitate compliance with the statutory requirement that all aircraft register prior to operation. The rule became effective on 21 December 2015 and all small unmanned aircraft had to be registered by 19 February 2016. Interestingly, the registration rule specifically excludes paper airplanes, Frisbees, and other flying objects, as these are not considered to be true unmanned aircraft. The FAA published its small UAS Notice of Proposed Rulemaking (NPRM) in February 2015, and the public comment period closed in April 2015. The FAA conducted a thorough analysis of all the comments received and produced a Summary of Responses (SOR) before issuing its notice of final rulemaking for small UAS. The FAA published an amendment to 14 CFR–Part 107 to allow the operation of small unmanned aircraft systems in the US NAS on 28 June 2016 and it became effective on 29 August 2016. The rule is titled, Operation and Certification of Small Unmanned Aircraft Systems. Along with the rule, the FAA published Advisory Circular 107-2, which provides guidance on remote pilot certification, remotely piloted aircraft (RPA) registration and marking, RPA airworthiness, and operation of small UAS in the US NAS.

Under the provisions of Part 107, small UAS can only be operated in visual line of sight (VLOS) of the remote pilot, i.e., the operator must be able to see the small unmanned aircraft at all times during flight. Part 107 allows for the operation of small unmanned aircraft in controlled airspace, providing the operator has prior authorization from Air traffic Control (ATC). The operation of small unmanned aircraft comes with a number of restrictions, which includes the requirement to operate them no faster that 87 knots (100 miles per hour) and they must be flown below 400 feet.

Part 107 provides the opportunity for operators of small UAS to deviate from the standard operating conditions imposed by the rule by allowing them to apply for a Certificate of Waiver (CoW), providing the operation can be safely conducted. Such deviations include, operation over people, operation in controlled airspace, operations at night and operation of multiple small UAS.

UK unmanned aviation regulations

Again, the other major aviation regulatory authority that did so much for manned aviation regulations does not yet have any regulations for UAS operations. However, while the UK CAA does not have any specific UAS regulations, it does have an excellent guidance document in the form of Civil Aviation Publication (CAP) 722, Unmanned Aircraft System Operations in UK Airspace—Guidance. This document provides guidance to persons who are involved in all aspects of the development of UAS on how to identify the route to certification, outline the methods by which permission for aerial work may be obtained and to ensure that all requirements are met by the UAS industry. The document highlights the safety requirements that an operator has to comply with before he/she is allowed to operate an unmanned aircraft in UK airspace.

Those intending to fly an unmanned aircraft below 20kg in weight in the UK are to be aware of the rules that are in place to keep everyone safe. Put simply, this means that an operator must:

- make sure he/she can see the unmanned aircraft at all times and doesn't fly higher than 400 ft;
- always keep the unmanned aircraft away from other aircraft, helicopters, airports and airfields;
- use common sense and fly safely; operators could be prosecuted if they don't comply.

Additionally, unmanned aircraft fitted with cameras must not be flown:

- within 50m of people, vehicles, buildings or other structures;
- over congested areas or large gatherings such as concerts and sports events.

Unmanned aircraft larger than 20kg must be able to automatically sense other aircraft and take corrective action to avoid them.

UAS regulations around the globe

A number of States have developed UAS regulations that allow access to airspace in their respective countries. While this access is not unfettered, it is more open than that allowed in the US. These nations include:

- Australia
- New Zealand
- Japan
- Brazil
- Mexico
- Sweden
- France and
- Czech Republic.

Canada and the UK, while not yet having any regulations, allow a certain amount of leniency in their use.

Australian UAS regulations

Australia was the first State to publish regulations for UAS. Mal Walker, a Flying Operations Inspector with CASA from 1996 to 2007, led a team that carried out a review of the legislation governing the operation of unmanned aircraft. He identified a lack of consistent legislation as an impediment to the progressive integration of these aircraft into civil airspace. Under Mal's guidance, the team developed CASR Part 101, which was published in 2002. With the publishing of this legislation, Australia became the first country in the world with an official UAS regulation. Interestingly, CASA anticipated an increase in civil operations of UAS and assumed that other States would quickly follow suit in developing their own UAS regulations and in 2004 commenced a project to conduct a post-implementation review of CASR Part 101. Unfortunately, the anticipation that the FAA and EASA would develop their regulations and thus CASA would be able to learn from them and refine its own legislation fell flat. The project was eventually cancelled in 2012. In July 2011, another project commenced to review CASR Part 101 and the associated guidance material relating to UAS and resulted in NPRM 1309OS in May 2014. The NPRM proposed a number of amendments to CASR Part 101, with particular reference to the establishment of a revised risk-based framework for regulating UAS operations. A key part of the proposed amendment acknowledged the existence of a "low-risk" class of UAS operations, which were determined as small unmanned aircraft with a gross weight of 2kg and below while they are being operated under a strict set of standard UAS operating conditions as defined and discussed in the NPRM. For these UAS operations, CASA proposed that the requirements for a UAV Controller's Certificate or an Unmanned Aircraft System Operator's Certificate (UOC) would not apply. Unmanned aircraft with a gross weight above 2kg in all operating conditions, and all unmanned aircraft operating outside of the standard operating conditions, would still require an operational approval. The amendment also proposed a number of changes to:

- update the current terminology used within CASR Part 101 to bring it in line with the latest terminology used by ICAO as found in Annex 2 to the Convention on International Civil Aviation—Rules of the Air;
- clarify the current requirements for training and certification;
- remove redundant requirements; and
- simplify the process for approval.

On March 30, 2016, CASR Part 101 was finally amended to align with ICAO terminology, in particular by replacing the term UAV with RPA. Key outcomes included simplified regulatory requirements for lower risk RPA operations and an allowance for more detailed operational matters to be dealt with in a Manual of Standards, providing greater flexibility and responsiveness in a rapidly evolving area. More specifically, the regulation established a set of standard operating conditions for RPA, categorizations for RPA according to weight or, in the case of airships, envelope capacity, and introduced the concept of "excluded RPA" to represent RPA operations considered to be lower risk, as determined by RPA category and operational use. Excluded RPA have reduced regulatory requirements, such as not needing an operator's certificate or a remote pilot license (RePL). The regulation also now permits private landowners to carry out some commercial-like operations on their own land under the "standard RPA operating conditions" without requiring them to hold a UOC or a RePL, if using an RPA weighing up to 25kg, provided that none of the parties involved receive remuneration. For RPA weighing between 25kg and 150kg, the operator needs to hold a RePL in the category of aircraft being flown. The regulation requires a person operating, or conducting operations using, a very small RPA for hire or reward to notify CASA rather than being required to obtain a UOC and RePL. The regulation makes it an offense for a person to operate a very small RPA for hire or reward without notifying CASA and also allows CASA to establish and maintain a database of information that relates to these notifications. The regulation inserts new definitions into Part 1 of the CASR Dictionary that align with ICAO definitions. Autonomous flight is prohibited under the amendments until such time as suitable regulations can be developed by CASA. However there is scope for autonomous flight to be approved by CASA on a case-by-case basis in the meantime. The regulation broadens the eligibility for a RePL by not specifically requiring an Aeronautical Radio Operator's Certificate (AROC), enabling the holder of an equivalent qualification to meet the required standards in respect of radio communications.

Indonesia UAS regulations

In 2015, the Indonesian Ministry of Transportation's Directorate General of Civil Aviation (DGCA) published a rule that regulates the usage of UAS in Indonesian airspace. The rule states that unmanned aircraft should not fly above an altitude of 150m and should not fly inside restricted or prohibited areas, and areas around airports. Unmanned aircraft that require altitudes higher than 150m require a written authorization from DGCA. Additionally, unmanned aircraft equipped with imaging equipment should not fly within 500m from the border of restricted or prohibited areas, and if the unmanned aircraft is involved in imaging activities, the operator should have written permission from the local government. Special unmanned aircraft equipped with farming equipment such as seed spreaders or insecticide sprayers should only operate in farmland, and should not operate within 500m of housing areas.

New Zealand UAS regulations

On September 24, 2015, the CAA of New Zealand released the following Civil Aviation Rules:

• Part 101—Gyrogliders and Parasails, Unmanned Aircraft (including Balloons), Kites, and Rockets—Operating Rules; and
• Part 102—Unmanned Aircraft Operator Certification.

Part 101 addresses the immediate safety risks associated with the use of RPAS, and moves New Zealand toward compliance with international Standards and Recommended Practices (SARPs) with regard to RPAS. Part 101 only applies to RPA of 25kg and under that can fully comply with the rules in Part 101. To operate any RPA above this weight, and for operations that cannot comply with Part 101, the operator must be certificated under Part 102. The objective of Part 102 is to introduce a new rule to address the immediate safety risks associated with the use of UAS, and to take steps to achieve compliance with international SARPs with regard to unmanned aircraft. The NZ CAA complimented its regulations with a comprehensive suite of Advisory Circulars (ACs) that contain information about standards, practices and procedures that the CAA NZ has found to be an acceptable means of compliance with the associated rule, in the form of:

• AC 101–1: Remotely Piloted Aircraft Systems (RPAS) under 25 kg— Operating in compliance with Part 101 Rules;
• AC 102–1: Unmanned Aircraft—Operator Certification.

Sweden UAS regulations

The Swedish Transport Agency (TSFS) published its regulations on UAS on November 5, 2009 (TSFS, 2009). The regulation is applicable to the design, manufacture, modification, maintenance and operational activities with civil UAS within Sweden that are not covered by EASA regulations. The regulations also apply to UAS used for testing or research, commercial purposes for hire and reward, professional occupation or similar activities not considered as recreation, and flight BVLOS.

UAS activities are subdivided into the following four categories:

- Category 1A: Unmanned aircraft with maximum take-off weight of less than or equal to 1.5 kg, which develops a maximum kinetic energy of 150 Joules and is flown only within the visual line of sight of the pilot.
- Category 1B: Unmanned aircraft with maximum take-off weight of more than 1.5kg but less than or equal to 7 kg, which develops a maximum kinetic energy of 1,000 Joules and is flown only within the visual line of sight of the pilot.
- Category 2: Unmanned aircraft with maximum take-off weight of more than 7kg which is flown only within the visual line of sight of the pilot.
- Category 3: Unmanned aircraft which is certified to fly and be controlled beyond the visual line of sight of the pilot.

The road to certification

As discussed earlier, no regulatory authority has yet developed any airworthiness regulations or design standards for UAS. Large drones are operated by the military and while they are built to military requirements, they do not meet civil aviation design standards. Small drones generally do not meet any standards at all and are usually manufactured using COTS components, which can be bought from any electronics store.

The main reason that there aren't any RPAS design standards available is because regulators don't yet know what such standards look like. Further, if there aren't any standards, it is impossible to write regulations to mandate the standards' use. In order to achieve a suitable set of design standards, regulators are going to have to work very closely with industry. Manufacturers such as Northrop Grumman, General Atomics and Boeing Insitu are already building unmanned aircraft that could potentially be certificated by civil regulators, if they had a mechanism to do so. This process of the regulator issuing a design standard or specification and then industry demonstrating a better way of doing things, which then amends the regulator's approach, is not new. A very good example of this occurred in 1930 in the UK when the Air Ministry issued AM Specification F7/30—for a new and modern fighter which had to be capable of speeds of 250mph and had to be able to carry four machine guns. Of the seven designs tendered, only one aircraft was accepted as being suitable, the Gloster Gladiator biplane. A number of the other designs came close, but all had some deficiencies. One of the other designers embarked on a series of "cleaned-up" designs in an attempt to show the Air Ministry that he could meet the specification. In July 1934, he submitted a new aircraft design to the Air Ministry, which had a retractable undercarriage and a smaller wingspan, but again, it was rejected. But our intrepid designer continued to persevere, and in December 1934 he presented yet another design. This time the Air

Ministry accepted the design as being superior and issued contract AM 361140/34 and provided £10,000 for the construction of his improved F7/30 design. Then on January 3, 1935, the Air Ministry formalized the contract and issued a whole new specification, F10/35, which was written around that specific aircraft. The young aircraft designer was R.J. Mitchell and the aircraft of course was the Spitfire. So, if aircraft designers keep putting their experimental RPAS to the regulators, eventually there might be a "Light-Bulb" moment and everybody will know what a type certificated RPAS looks like.

References

A Brief History of the FAA, www.faa.gov.

CAP 789 Requirements and Guidance Material for Operators 18 February 2011 (Supersedes CAP 768, which supersedes CAP 360 Part 1 "The Air Operator's Certificate" (AOC)).

Chaplin, JC "Safety Regulation – The First 100 Years"(2011) *Journal of Aeronautical History*.

Edwards, Ronald "British Air Transport in the Seventies – A Report of the Committee of Inquiry into Civil Air Transport" May 1969.

FAA Lessons Learned—Midair Collision between a Trans World Airlines Lockheed 1049A and a United Airlines Douglas DC-7, Grand Canyon, Arizona, June 30, 1956, http://lessonslearned.faa.gov/ll_main.cfm?TabID=1&LLID=50, accessed August 11, 2016.

Friedman, Herbert M and Friedman, Ada Kera "The Legacy of the Rockne Crash" *Aeroplane Magazine* (Article provided by the University of Notre Dame Archives, posted on the website "Reflections from the Dome") May 2001.

Glass, Andrew "Sen. Bronson Cutting dies in plane crash, May 6, 1935" May 6, 2015, www.politico.com/story/2015/05/sen-bronson-cutting-died-may-6-1935-117639#ixzz4HNbf11Ku, accessed August 11, 2016.

Gorell Ronald B. "Memorandum by the Secretary of State for Air on the Report of the Committee on Control of Private Flying and other Civil Aviation together with the Report of the Committee and the Appendices thereto" 1934.

Sykes, Sir Frederick *From Many Angles: An Autobiography* G.G Harrap, London,1943.

The Swedish Transport Agency's Regulations on Unmanned Aircraft Systems (UAS) TSFS 2009:88, November 5, 2009.

Towey, Barrie *Jet Airliners of the World 1949–2007* Air-Britain, 2007.

4 Global harmonization
International Civil Aviation
Organization

It always seems impossible until it is done.

Nelson Mandela

While the first successful flight of a manned fixed wing aircraft occurred in 1903, it took another eleven years for the first passenger-carrying fixed wing flight, which occurred on January 1, 1914 between St. Petersburg, Florida and Tampa, Florida across Tampa Bay using a Benoist XIV flying boat (Glines, 1997). It then took a further two years before we saw the first fixed wing airline in operation. This occurred in Europe with the "Aircraft Transport and Travel" (AT&T) airline, formed by George Holt Thomas in 1916. AT&T used a fleet of former military Airco DH.4A biplanes that had been modified to carry two passengers in the fuselage; it operated flights between Folkestone in the UK and Ghent in Belgium.

The Deutsche Luftschiffahrts-Aktiengesellschaft (DELAG) was officially the world's first airline, founded on November 16, 1909; however, it operated Zeppelin airships, not the fixed wing aircraft that we associate with an airline of today (see Airships.net, 2010). The five oldest airlines that still exist today and still operate fixed wing aircraft are Netherlands' KLM (founded October 7, 1919), Colombia's Avianca (founded December 5, 1919), Australia's Qantas (founded November 16, 1920), Russian Aeroflot (founded February 9, 1923), and the Czech Republic's Czech Airlines (founded October 6, 1923).

As early as 1919, questions were being asked about air jurisdiction, which is the fundamental right of a sovereign state to regulate the use of its airspace and enforce its own aviation law. A Convention held in Paris in 1919, under the auspices of the International Commission for Air Navigation (ICAN), sought to answer this question and it was decided that each nation has absolute sovereignty over the airspace overlying its territories and waters. The following principles governed the drafting of the Convention (Paris Convention, 1919):

1. Each nation has absolute sovereignty over the airspace overlying its territories and waters. A nation, therefore, has the right to deny entry and regulate flights (both foreign and domestic) into and through its airspace.

2. Each nation should apply its airspace rules equally to its own and foreign aircraft operating within that airspace, and make rules such that its sovereignty and security are respected while affording as much freedom of passage as possible to its own and other signatories' aircraft.
3. Aircraft of contracting states are to be treated equally in the eyes of each nation's law.
4. Aircraft must be registered to a state, and they possess the nationality of the state in which they are registered.

The treaty was signed on October 13, 1919 by 26 nations but it was ratified by only 11, including Persia, which interestingly had not signed the treaty in the first place. The treaty came into force in 1922.

On June 15, 1929, a supplement to the Paris Convention of 1919, known as the Protocol, was signed in Paris and made reference to "pilotless aircraft" in a subparagraph of Article 15 as follows:

> No aircraft of a contracting State capable of being flown without a pilot shall, except by special authorization, fly without a pilot over the territory of another contracting State.
>
> (Paris Convention Protocol, 1929)

The ICAN had its first meeting in Berlin in 1903 and continued to operate until 1945. On December 7, 1944, a Convention on International Civil Aviation was held in Chicago, Illinois, which later became known as the Chicago Convention, and was signed by 52 States. This Convention required ratification by 26 States before being accepted. Under the terms of the Convention, on June 6, 1945, a Provisional International Civil Aviation Organization (PICAO) was established and ICAN was disestablished. On March 5, 1947 the 26th ratification of the Convention was received. Consequently, ICAO was established on April 4, 1947 and replaced PICAO.

When the Chicago Convention replaced the Paris Convention, Article 8 of the Chicago Convention also replaced Article 15 the 1929 Protocol and the term "pilotless aircraft" took on the definition:

> No aircraft capable of being flown without a pilot shall be flown without a pilot over the territory of a contracting State without special authorization by that State and in accordance with the terms of such authorization. Each contracting State undertakes to insure that the flight of such aircraft without a pilot in regions open to civil aircraft shall be so controlled as to obviate danger to civil aircraft.
>
> (Chicago Convention, 1944)

To understand the implications of Article 8 and its incorporation into the Chicago Convention of 1944, the intent of the original drafters of the legislation must be considered. Remotely controlled aircraft were already in existence at the time of World War I, operated by both civil and military entities. "Aircraft flown without a pilot" therefore refers to the situation where there is no pilot onboard the aircraft.

The Eleventh Air Navigation Conference, held in Montréal, Canada from September 22 to October 3, 2003, endorsed the global ATM operational concept, which contains the following text:

> [a]n unmanned aerial vehicle is a pilotless aircraft, in the sense of Article 8 of the Convention on International Civil Aviation, which is flown without a pilot in-command on-board and is either remotely and fully controlled from another place (ground, another aircraft, space) or programmed and fully autonomous.

Consequently, any unmanned aircraft is a "pilotless aircraft," which is consistent with the intent of Article 8. Emphasis was placed on the significance of the provision that aircraft flown without a pilot onboard "should be so controlled as to obviate danger to civil aircraft," indicating that the legislation drafters recognized that "pilotless aircraft" must have a measure of control applied to them in relation to a so-called "due regard" obligation, similar to that of State (military) aircraft.

On April 12, 2005, during the first meeting of its 169th Session, the Air Navigation Commission (ANC) requested the Secretary General to consult selected States and international organizations with respect to present and foreseen international civil UAV activities in civil airspace; procedures to obviate danger to civil aircraft posed by UAVs operated as State aircraft; and procedures that might be in place for the issuance of special operating authorizations for international civil UAV operations. Subsequently, a first ICAO exploratory meeting on UAVs was held in Montréal, Canada on May 23 and 24, 2006. Its objective was to determine the potential role of ICAO in UAV regulatory development work. The meeting agreed that, although there would eventually be a wide range of technical and performance specifications and standards, only a portion of those would need to become ICAO SARPs. It was also determined that ICAO was not the most suitable body to lead the effort to develop such specifications. However, it was agreed that there was a need for harmonization of terms, strategies and principles with respect to the regulatory framework and that ICAO should act as a focal point.

A second informal ICAO meeting held in Palm Coast, Florida on January 11 and 12, 2007, concluded that work on technical specifications for UAV operations was well underway within both RTCA Inc. (formerly known as the Radio Technical Commission for Aeronautics) and the European Organization for Civil Aviation Equipment (EUROCAE) and was being adequately coordinated through a joint committee of their two working groups. The main issue for ICAO was, therefore, related to the need to ensure safety and uniformity in international civil aviation operations. Consequently, it was agreed that there was no specific need for ICAO to develop any new

SARPs at that early stage. However, the general consensus was that there was a need to harmonize concepts and terminology. The meeting agreed that ICAO should coordinate the development of an advisory document that would guide the regulatory evolution. ICAO guidance documents, while non-binding, can be used as the basis for the development of regulations by the various States. ICAO anticipated that as States developed their own regulations and/or guidance material, this would be fed back to ICAO for inclusion in the ICAO guidance document. The ICAO material in conjunction with State material would then complement each other and the document would then serve as the basis for achieving consensus in the later development of SARPs.

The meeting in 2007 also recommended that the development of SARPs should be undertaken in a systematic and coordinated fashion. Further, since it had been recognized that there were other technical organizations and regulatory authorities that had been working with unmanned aircraft a lot longer than ICAO, these groups probably had a better grasp of the emerging technologies than did ICAO. Consequently, it was decided that ICAO should seek to harmonize with these other bodies at the earliest stage. The meeting also suggested that the term UAV should be dropped in favor of the term UAS, in line with RTCA and EUROCAE. Finally, it was concluded that the best position for ICAO was to take an international leadership role and serve as the focal point for global interoperability and harmonization. ICAO's prime role would be the development of UAS SARPs. Moreover, it was deemed that ICAO also had the capacity to contribute to the development of technical specifications being undertaken by other technical organizations and to identify communication requirements for UAS activities.

Unmanned aircraft systems study group

In order to support ICAO targets, the ANC, at the Second Meeting of its 175th Session on April 19, 2007, approved the establishment of the Unmanned Aircraft Systems Study Group (UASSG), where the following Terms of Reference and Work Program were defined as:

UASSG Terms of Reference:

> In light of rapid technological advances, to assist the Secretariat in coordinating the development of ICAO Standards and Recommended Practices (SARPs), Procedures and Guidance material for civil unmanned aircraft systems (UAS), to support a safe, secure and efficient integration of UAS into non-segregated airspace and aerodromes.
>
> (ICAA Doc 10019, 2015)

UASSG Work Program (ICAA Doc 10019, 2015):

* serve as the focal point and coordinator of all ICAO UAS related work, with the aim of ensuring global interoperability and harmonization;

- develop a UAS regulatory concept and associated guidance material to support and guide the regulatory process;
- review ICAO SARPs, propose amendments and coordinate the development of UAS SARPs with other ICAO bodies;
- contribute to the development of technical specifications by other bodies (e.g. terms, concepts), as requested; and
- coordinate with the ICAO Aeronautical Communications Panel (ACP), as needed, to support development of a common position on bandwidth and frequency spectrum requirements for command and control of UAS for the International Telecommunication Union (ITU)/World Radiocommunication Conference (WRC) negotiations.

The UASSG was tasked, *inter alia*, with developing a regulatory concept and associated guidance material to support and guide the regulatory process for UAS. The UASSG took the approach that an unmanned aircraft is not actually "unmanned" but rather is piloted, albeit from a remote location and thus introduced the term "remotely piloted." The use of this term demonstrated that there was actually a pilot in the loop. The result of looking at the system in this manner set the scene that only unmanned aircraft that are remotely piloted could be integrated alongside manned aircraft in non-segregated airspace and at aerodromes. The first deliverable of the UASSG, published in March 2011, was Circular 328 (Cir 328)—Unmanned Aircraft Systems (Cary and Coyne, 2012). While the Circular introduced the terms RPAS, RPA, remote pilot station (RPS) and remote pilot, the definitions were further refined at a later stage, which will be discussed later in this chapter. It also provided an overview of a number of issues that would have to be addressed in the SARPs to ensure that RPAS would be compliant with the provisions of the Chicago Convention. In March 2012, the UASSG achieved its second deliverable with amendments to the ICAO Annex 2—Rules of the Air and Annex 7—Aircraft Nationality and Registration Marks.

In March 2014, the UASSG achieved its final deliverable, the Manual on RPAS, which contains material that is recommended for use by ICAO contracting States and Regional Safety Oversight Organizations when establishing the regulatory framework for RPAS certification and subsequent operations. ICAO published this Manual as Doc 10019 in March 2015.

RPAS Manual

The Manual on RPAS was the major piece of work produced by the UASSG. The purpose of the manual is to provide guidance on the technical and operational issues applicable to the integration of RPA in non-segregated airspace and at aerodromes.

As discussed earlier, the UASSG decided to concentrate its efforts on UAS that are remotely piloted, i.e. RPAS. The decision was also taken that State (military) aircraft, autonomous unmanned aircraft, which includes unmanned free balloons or other types of aircraft which cannot be managed on a real-time basis during flight, and model aircraft would be excluded from the Work Program.

The UASSG built on the work of Cir 328, expanded the previous definitions recorded in that document and established a new set of definitions and terms that are unique to this new technology that includes:

- Remotely Piloted Aircraft Systems (RPAS)—A remotely piloted aircraft, its associated remote pilot station(s), the required C2 links and any other components as specified in the type design.
- Remotely Piloted Aircraft (RPA)—An unmanned aircraft that is piloted from a remote pilot station.
- Remote Pilot Station (RPS)—The component of the RPAS containing the equipment used to pilot the RPA.
- Remote crewmember—A crewmember charged with duties essential to the operation of a RPAS during a flight duty period.
- Remote pilot—A person charged by the operator with duties essential to the operation of a RPA and who manipulates the flight controls, as appropriate, during flight time.
- RPAS operator certificate (ROC)—A certificate authorizing an operator to carry out specified RPAS operations.

RPAS

When we speak about an aircraft, we are really talking about a system of components all flying in unison. These components include the airframe, the engines, the cockpit with all the instrumentation, including the pilots, the communications between the pilots, air traffic control and other airspace users, i.e. other aircraft. Similarly, with unmanned aircraft, we also need to talk about the whole system, i.e. the RPAS. This consists of the airframe (RPA), the cockpit (RPS), which is now located separate from the airframe and includes the pilots and the communications between the pilots, air traffic control and other airspace users. We then add an additional crucial component, the C2 data-link so that the RPA and RPS are electronically linked to manage the flight. There is also a bunch of other ancillary equipment or components, such as a launch and recovery system, cameras, video etc. that will be included as required and that ensure safe operation.

The operation of a RPAS is quite different from that of a manned aircraft. For example, international operations of a manned aircraft are considered to be those in which the aircraft crosses an international border or operates over the high seas. With RPAS, the RPA and/or the RPS can be operated in a country other than the country of the operator. What does this mean? An aircraft designer in the USA decides to build RPAs, while an electronics organization in Australia decides to build RPSs. The RPA builder wants to fly his RPA from Los Angeles (LA) to Auckland, but the problem is, the RPA doesn't have a cockpit (RPS) and there are no pilots. The RPS builder decides to locate RPSs in LA, Hawaii, Sydney and Auckland. The RPS builder also has remote pilots. Problem is, no RPA. The RPA builder and the RPS builder get together and under a Memorandum of Agreement (MOA), the RPA builder includes the RPSs on the Type Certificate Data Sheet (TCDS) and gets a TC from the FAA for an RPAS. The RPA can then be launched from LA by RPS1, handed over to RPS2 in Hawaii, handed over to RPS3 in Sydney, and then finally handed over to RPS4 in Auckland for landing. Job done. The TC process will be expanded on later in this chapter.

With manned aviation operations, we generally talk about operations under the IFR or VFR, which then determine the requirements for equipment, operations and responsibilities. With RPAS operations, it is more appropriate to talk about operations in VLOS or BVLOS of the remote pilot. With VLOS, the remote pilot or RPA observer must maintain direct unaided visual contact with the RPA at all times during the operation. BVLOS operations are when neither the remote pilot nor the RPA observer can maintain direct unaided visual contact with the RPA. Minimum equipment requirements to support BVLOS operations increase significantly as the range and complexity of such operations increase, as does the cost involved in ensuring the robustness of the C2 data-link. The ability to detect and avoid conflicting traffic or obstacles and take appropriate action to avoid them is essential.

To fly an RPA internationally, an operator must obtain a special authorization from any country intending to be overflown. This can be achieved by countries establishing formal agreements in the form of bilateral or multilateral memoranda of understanding (MoU) to adopt simpler procedures. It is also imperative that before any operations of RPA occur over the high seas, the appropriate air traffic services (ATS) authority(ies) are contacted and programs and strategies are coordinated. During the RPA flight planning phase of international flights, the remote pilot should seek prior authorization from countries that may be overflown due to emergencies or unexpected meteorological conditions that may require the remote pilot to fly alternate routes or even to have to land at an alternate aerodrome that is located in another country.

Type certification and airworthiness approvals

ICAO requires every aircraft that will operate in international airspace to have a CofA (ICAO Doc 9760, 2014). This is irrespective of whether the aircraft is manned or unmanned. Consequently, existing processes and procedures that are applied to manned aircraft type design approvals, such as type certification, production approval, continuing airworthiness and modifications/alterations of aeronautical products are also applicable to RPAS. As discussed previously, this will of course pose a new set of challenges for countries as the RPAS could be distributed between a number of countries, i.e. there could be several RPSs distributed around the globe that are used to control an RPA as it flies on an international route. This may involve different countries having different responsibilities in relation to the different parts of the RPAS. It will also impact on which country has the responsibility for States of Design, Manufacture, Registry and the Operator and their respective oversight requirements.

The RPA is the aircraft component of the RPAS and ICAO has determined that the RPA must have type design approval in order to operate internationally. This would follow processes and procedures similar to a manned aircraft and the RPA will be required to have a type design approval in the form of an RPA TC, which would be issued to an RPA TC holder. An RPA TC holder is a responsible entity, which can be an organization or a person. The TC is issued when the responsible entity has achieved compliance with an appropriate and agreed type certification basis. The certification basis would include applicable requirements similar to manned aircraft in all appropriate areas of design and construction, for example, structures and materials, electrical and mechanical systems, propulsion and fuel systems, and flight testing. Due to the nature of the RPA requiring one or more RPS(s) to be able to operate, i.e. to make it into an RPAS, the design approval requirements will need to be expanded from the RPA itself to include the RPS(s), the C2 data-link(s) and any other components of the system to enable safe operation from take-off to landing. When the RPAS conforms to an approved type design and is compliant with all other continuing airworthiness requirements, such as maintenance actions and configuration control, the RPA part of the RPAS is considered to be airworthy and thus safe to fly.

The remote pilot must be able to manage the flight of the RPA from the RPS on a real-time basis through use of the C2 data-link. Consequently, it is imperative that the C2 data-link must be part of the TC process for the RPAS. This adds a further layer of complexity as the C2 data-link could be procured from a service provider under contract to the RPAS operator. Notwithstanding, safe integration of the C2 data-link into the overall design of the RPAS is the responsibility of the RPA TC holder.

While the RPS must be approved, it doesn't necessarily have to have its own TC. If we think of the RPS being akin to an engine or propeller on a manned aircraft, which can either have their own TC or be part of the aircraft TC, the same can be achieved with the RPS and indeed any other component of the RPAS. The RPA TC holder needs to demonstrate that the integration of all the various components, engine, propeller, RPS etc. has been achieved. As with manned aviation, the FAA Technical Standard Order (TSO) process can be applied to the various components of the RPAS to reduce the burden of verification.

An aeronautical product, as defined in the ICAO Airworthiness Manual, is any aircraft, aircraft engine, aircraft propeller or a part thereof, including any associated computer system and computing software. The C2 data-link is a telecommunications link over which data is transmitted. It is essentially a circuit that electronically connects the RPA to the RPS and is an integral part of the RPAS. Consequently, it is not an aeronautical product and cannot be independently type certificated. However, there is nothing stopping the C2 data-link from being part of the overall type design of the RPAS. One of the other issues with a data-link is that it needs to be protected from being hacked or spoofed by the nefarious activities of some malcontent. It also has to be secure from all other forms of interference, whether intentional or unintentional.

RPA registration

In accordance with the Chicago Convention, all civil aircraft must be registered with a national aviation authority and all aircraft that are engaged in international operations must display its registration marks prominently on the aircraft. Each national aviation authority allocates a unique alphanumeric code to identify the aircraft, which then indicates the country of registration of the aircraft. For example, aircraft registered in the USA start with the letter "N," aircraft registered in the UK start with the letter "G" and in Australia, "VH." This is fine when the aircraft is large enough to be able to paint the registration marks on. However, the sizes and shapes of RPA can differ significantly from those of current manned aircraft. The UASSG identified this problem early in its work program and amended Annex 7—*Aircraft Nationality and Registration Marks*, to accommodate these differences. This amendment gives the State of Registry of the RPA, the authority to determine the location and measurement of registration marks.

RPASP

The UASSG, while it accomplished some very good work, worked in a different framework than a recognized Panel, which created some issues for ICAO. Historically, study groups are usually small groups of approximately six to ten technical experts managed by the Secretariat, which are tasked to investigate a particular technical issue and make recommendations on whether a Panel needs to be established under the direction of the ANC to develop SARPs. However, the work that the study group was doing gathered momentum quickly and the study group grew into the UASSG. The first meeting of the UASSG in April 2008 had 30 attendees and the last meeting in July 2015 had 64 attendees; not your typical study group. At the Second Meeting of the AIR Navigation Council's 196th Session on May 6, 2014, it was agreed to establish a Remotely Piloted Aircraft Systems Panel (RPASP). The UASSG had its final meeting in July 2014. The RPAS Panel was then established and had its first meeting in November 2014 and, under the direction of the ANC, it was tasked with taking up where the UASSG had left off and to continue to serve as the focal point and coordinator of all ICAO RPAS-related work, with the aim of ensuring global interoperability and harmonization. The RPASP was tasked with developing a RPAS regulatory concept and associated guidance material to support and guide the regulatory process. The RPASP is currently developing SARPs for the licensing, airworthiness, operations and equipment provisions of RPAS.

Detect and avoid

In manned aviation the pilot can generally see out of the window and is therefore able to see other traffic and maneuver his/her aircraft to avoid a conflict or a mid-air crash. This is known as "see and avoid" and is a typical human function. More sophisticated aircraft and certainly passenger-carrying aircraft are fitted with a variety of computer hardware and clever software technology to assist in avoiding conflicts with other traffic. Such equipment includes automatic dependent surveillance-broadcast (ADS–B). An aircraft fitted with this equipment can determine its position using GPS and then it broadcasts information, such as the aircraft's position, altitude and velocity via a data-link. Thus the aircraft can be tracked by air traffic control on the ground and other aircraft can receive the data that provides situational awareness and allows the aircraft to maintain separation. This is known as "sense and avoid" and is a typical computer function.

Humans surpass computers in many areas, such as:

- the ability to detect small amounts of visual and acoustic energy;
- the ability to perceive patterns of light or sound;
- the ability to improvise and use flexible procedures;
- the ability to store very large amounts of information for long periods and to recall relevant facts at the appropriate time;
- the ability to reason inductively (bottom up approach); and
- the ability to exercise judgment.

Computers, on the other hand, surpass humans in the following areas:

- the ability to respond quickly to control signals;
- the ability to perform repetitive, routine tasks;
- the ability to store information briefly and then to erase it completely;
- the ability to reason deductively (top down approach); and
- the ability to handle highly complex operations, i.e. to do many different things at once.

The UASSG took the view that for RPAS operations, both the "see and avoid" and "sense and avoid" functions are required. This combination of both the human and computer attributes is known as "detect and avoid" (DAA). ICAO defines DAA as the "capability to see, sense or detect conflicting traffic or other hazards and take the appropriate action." This capability is viewed as the minimum requirement to ensure the safe operation of an RPA flight and to enable full integration in all airspace classes with all airspace users. A solution for DAA has not yet been found, so for the RPAS world, it is the "Holy Grail" and whenever that nut is cracked, we will see an explosion in the RPAS world. Further, whoever finds the solution will be sitting on a gold mine.

So what does a DAA system look like? It is envisaged that a combination of technology and procedures will be required to give the same capabilities as those that pilots of manned aircraft have, using one or more senses (e.g. vision, hearing, touch) and associated cognitive processes. RPAS of the future will probably have a variety of different sensors to detect different hazards. The output from these sensors will then be fed to systems that will ensure that the RPA avoids a conflict with other aircraft, persons on the ground or property. DAA systems may use more than one sensor to guarantee reliable hazard detection. This poses additional complexity, as these systems will have to be interoperable to ensure that an appropriate, coordinated avoidance action is taken when different hazards are present at the same time.

Hazard identification and risk analysis

RPA must be able to detect hazards and this can be achieved by either, "see and avoid" and/or "sense and avoid" techniques. Optical systems can provide the "see" function and these can take the form of "First Person View" (FPV), cameras including video, thermal imaging and light detection and ranging (LIDAR) systems. As discussed above, the "sense" function can be achieved using ADS-B, but could equally include radar, airborne collision avoidance systems (ACAS), FLARM (a portmanteau of "flight" and "alarm"), ground proximity warning systems (GPWS) or synthetic vision systems (SVS).

Hazards exist in all aviation activities. It is important to be able to identify those hazards, assess the risks associated with the hazards and develop appropriate mitigation strategies to minimize the risks. The ICAO Global Air Traffic Management Operational Concept (ICAO Doc 9854, 2005) identifies the need to limit the risk of collision to an acceptable level between an aircraft and the following hazards:

> other aircraft, terrain, weather, wake turbulence, incompatible airspace activity and, when the aircraft is on the ground, surface vehicles and other obstructions on the apron and maneuvering area.

The document also notes:

> for any hazard (i.e. any condition, event or circumstance that could induce an accident), a risk can be identified as the combination of the overall probability or frequency of occurrence of a harmful effect induced by the hazard, and the severity of that effect.
>
> (Doc 9854, 2005)

Both manned aircraft and RPA may encounter any of the above-mentioned hazards during operations. However, it is important to note that what might be a minor risk of the hazard for an RPA, may not be true for a manned aircraft encountering the same hazard in the same airspace and vice versa. When conducting a risk analysis, one must be cognizant of this and tailor the risk analysis appropriate to both the manned aircraft and the RPA. It is a good rule of thumb to conduct two separate risk analyses, one for the manned aircraft and one for the RPA.

C2 data-link

As discussed earlier the C2 data-link is a telecommunications link over which data is transmitted and electronically connects the RPA and the RPS. It provides the connection between the remote pilot and the RPA. It essentially takes the place of the control wires or data bus between the cockpit and the control surfaces found on manned aircraft. The prime purpose of the C2 data-link is to uplink commands to modify the behavior of the RPA, i.e. to fly the RPA, and to downlink information to indicate the position and status of the RPA, i.e. being the eyes of the remote pilot. The C2 data-link can also be used to relay voice and data communication between air traffic control and the remote pilot.

You don't have to be a rocket scientist to work out that the C2 data-link is the lifeline between the RPS and RPA. Lose the link and there goes the ability for the remote pilot to fly the RPA, leaving the RPA unmanned and uncontrolled. In fact, it could be said that the RPA is now flying autonomously, i.e. under its own control. The level of the hazard and the risks associated with loss of the C2 data-link will be determined by where the RPA is in its phase of flight, e.g. if it is flying in segregated airspace with no other aircraft in the vicinity, the risks are a lot lower than if the RPA is in the final stages of a landing. If the C2 data-link is lost, there should be a lost link procedure in place and it should be initiated once the C2 link cannot be used to control the RPA. Although the C2 data-link can be lost for short or long periods of time, the RPAS design must be such that loss of the C2 data-link should not result in a hazardous or catastrophic event (e.g. collision with another aircraft or uncontrolled collision with the ground or obstacle). Some features that could be incorporated into the design of the RPAS to ensure that such an event does not occur if the C2 data-link fails include:

- have the RPA return to home or to a pre-determined safe location;
- allow the RPA to complete its mission and land at a pre-determined safe location;
- have the RPA immediately land;
- have the RPA descend to a low altitude such that it cannot hit another aircraft and then fly out to sea and ditch; or
- have the RPA climb to altitude in an attempt to regain the C2 data-link.

Any of these options, or indeed any other scenario, will be the subject of intense scrutiny as to the airspace the RPA will be flying in, as well as the RPA altitude and speed, and an awareness of the segment of flight at all points on the flight plan where the failure could occur. All these will need to be taken into account when designing the software that will provide the commands to the onboard computer that is now flying the RPA. The selected option will also need to be cleared by air traffic control prior to the flight. Additionally, the remote pilot must advise air traffic control as soon as the C2 data-link is lost.

Air traffic control communications

Air traffic control communications are different from the C2 data-link communications. The requirements for RPA are the same as for manned aviation operating in the same airspace. This could be either very high frequency (VHF) voice or Controller-Pilot-Data-Link-Communications (CPDLC). This could also include telephone backup, providing it is approved by air traffic control beforehand. Communication between air traffic control and the remote pilot can be either via the RPA or via a broadcast, private or service provider network. Both methods have their advantages and disadvantages. Communication via the RPA is probably the easiest as it is transparent to air traffic control and is compatible with existing air traffic control communications with manned aircraft pilots across the globe. Its main disadvantage is that it may require more bandwidth for the C2 data-link to support the air traffic control voice and data relay between the RPA and the remote pilot. Communications via a network will be more complex and may require additional infrastructure or equipment in the air traffic control unit.

ATM integration

The majority of RPA operations around the globe are currently constrained to designated portions of the NAS known as segregated airspace, which is reserved for specific users, e.g. RPA. In fact, many RPA will never operate in anything but segregated airspace. To be able to fly in non-segregated airspace, RPA must be able to ensure appropriate separation with other aircraft and avoid mid-air collisions. While some RPA have been given access to non-segregated airspace, this has only been granted for limited operations and has required specific safety procedures.

Integration of RPA in non-segregated airspace will be a gradual process. There are a few regulatory and technical challenges to overcome before we see RPA mixing it up with manned aircraft. RPAS design standards still need to be developed, which includes standards for DAA as discussed earlier. Regulatory authorities need to develop appropriate regulations for certification. RPAs have to have the same level of flight safety when operating alongside manned aircraft.

It is not enough simply to talk about integration of RPA in non-segregated airspace, we must also add in the factor of whether we are talking about controlled or uncontrolled airspace. In order to operate in non-segregated uncontrolled airspace, the RPA will need to be able to interact with other airspace users, without impacting the safety or efficiency of existing flight operations. To be integrated into non-segregated controlled airspace, the RPA must be able to comply with existing ATM procedures. Further, depending on the class of airspace within which the RPA is operating, it will need to carry a transponder the same as manned aircraft. A transponder is an electronic device that makes it possible for air traffic control to identify the RPA's position and altitude.

There are a lot of different opinions about what constitutes an emergency with an RPA. For example, many people regard the loss of the C2 data-link as an emergency situation. Thus, if the C2 data-link is lost then the remote pilot should set the transponder code to 7700—Emergency. Alternatively, there are others who do not consider the loss of the C2 data-link as an emergency situation but rather a loss of communications. Thus the remote pilot may set the transponder code to 7600—Lost Communications. Then there are others who consider that the loss of the C2 data-link is neither an emergency situation nor a loss of communications. Let us look at the rationale for such thinking. In regards to emergency situations, providing the designer has built in sufficient safeguards that provide appropriate mitigation strategies, an emergency situation will not arise. So, there is no requirement to set the transponder code to 7700. Further, the use of code 7600 indicates a radio communication failure, which is inappropriate for loss of data-link. Notwithstanding the above, there is no definitive answer to the issue and a number of regulatory authorities are investigating a new non-discrete code for use by RPA to indicate loss of the C2 link.

The future

ICAO will continue to serve as the RPAS focal point to ensure global interoperability and harmonization. Through the RPAS Panel, it will continue the development of RPAS SARPs and will continue to revise and amend the RPAS Manual to keep it in line with the developing SARPs.

ICAO has also embarked on a schedule of training, starting with an RPAS workshop that will provide much-needed background knowledge to regulators on the procedures and guidance material contained in the RPASM, as well as on the ongoing work on the development of the SARPs.

References

Airships.net "DELAG: The World's First Airline" August 22, 2010, www.airships.net/delag-passenger-zeppelins, accessed May 15, 2016.

Cary, Leslie and Coyne, James *ICAO Unmanned Aircraft Systems (UAS), Circular 328. UAS Yearbook – UAS: The Global Perspective 2011–2012* Blyenburgh & Co., Paris, 2012.

Convention on International Civil Aviation held at Chicago (Chicago Convention), December 7, 1944, www.icao.int/publications/Documents/7300_orig.pdf, accessed August 15, 2016.

Convention Relating to the Regulation of Aerial Navigation (Paris Convention). League of Nations Treaty Series, October 13, 1919.

Glines, CV "Aviation History, St. Petersburg-Tampa Airboat Line: World's First Scheduled Airline Using Winged Aircraft" May 1997, https://info.aiaa.org/Regions/SE/Commercial%20Aviation%20HAS/2%20-%20Saint%20Petersburg-Tampa%20Airboat%20Line.pdf, accessed August 15, 2016.

ICAO Airworthiness Manual (Doc 9760) 3rd Edition, 2014, www.afeonline.com/shop/icao-doc-9760.html, accessed August 15, 2016.

ICAO Global Air Traffic Management Operational Concept (Doc 9854) 1st Edition, 2005, www.icao.int/Meetings/anconf12/Document%20Archive/9854_cons_en[1].pdf, accessed August 15, 2016.

ICAO Manual on Remotely Piloted Aircraft Systems (RPAS) (Doc 10019) 1st Edition, 2015, www4.icao.int/demo/pdf/rpas/10019_cons_en%20-%20Secured.pdf, accessed August 15, 2016.

Protocol to amend the Convention Relating to the Regulation of Aerial Navigation (1919 Paris Convention), June 15, 1929, www.icao.int/Meetings/LC36/Working%20Papers/LC%2036%20-%20WP%202-4.en.pdf, accessed August 15, 2016.

5 The good, the bad and the ugly

UAS applications and technology

> In the twenty-first century, the robot will take the place which slave labour occupied in ancient civilization.
>
> Nikola Tesla

Whether technology is good, bad or ugly is in the eyes of the beholder. Teenagers think that computers are good as they automatically correct their spelling and grammar mistakes while they are doing their homework. Senior citizens think that computers are bad because children have now lost the art of penmanship. Our older generation would rather receive a hand-written letter than an e-mail, whereas most teenagers have probably never written a letter. Teenagers think that texts on their phone look good and understand the words without any problems, whereas our parents' generation think that texts look ugly and cannot for the life of them make out the meaning.

While most new technology is usually created for the betterment of society, can any technology be considered good, bad or even ugly? According to case law, such as *MGM Studios* versus *Grokster*, the courts decided that technology is neither intrinsically good nor bad (*MGM* v. *Grokster*, 2005). This concept can be extrapolated to also include the fact that it cannot be ugly either. Further, those of us who grew up arguing whether Betamax was better that VHS or vice versa in the 1970s understand just how much technology invades our lives. So, it is not the technology, but rather it is the people operating the technology who are responsible for how it is used. For example, most of us would agree that mobile phones are neat pieces of technology and the ability to be able to text your friends or send photographs of your holiday snaps is good. However, doing so while you are driving is illegal and would thus be considered bad. Using it for cyber bullying is a good example of the ugly side of mobile phones.

The internet

The internet can be a wonderful tool. I have recently become engrossed in researching my family tree. I have joined a number of genealogy websites and I can search numerous databases quickly. I can check births, marriages and death records, or "hatches, matches and dispatches" as they have become known, back hundreds of years. I have researched back to the 1700s and discovered who my great, great grandparents were and where they lived in Ireland. I have also found the dates that my great grandmother left Ireland and arrived in Liverpool. I then get to discuss the stories with my friends and family, which makes for wonderful conversation. This can easily be regarded as a very positive, i.e. a good experience. However, the internet can take on a bad side when moderation goes out the door. If its use becomes excessive, compulsive or even addictive to a point where it interferes with a person's everyday life, it may be a problem. Children or teenagers playing online games for hours on end to the detriment of their schoolwork, sleep or social interaction can end up with psychological problems. A TV show aired on SBS on April 12, 2016, showed young teens in South Korea who were so addicted to gaming they could no longer distinguish between the real and imaginary worlds. The show followed them as they underwent intense rehabilitation. One teen said: "I feel like the game is controlling me, and when I lose it, I lose my temper" (Woo, 2016).

Several teenagers were sent to the National Centre for Youth Internet Addiction Treatment. It was set up by the South Korean government with the aim of tackling their addiction problems before it became too late. One youth admitted that he played games for 20 hours a day on weekends and 10 to 14 hours a day on school days (Seo, 2016).

In an article "How do you raise healthy kids in the age of technology? Here's how to manage their screen time to avoid negative impacts," Megan Blandford (2016) cites a recent Royal Children's Hospital Australian Child Health Poll, where 58 percent of parents said that their top health concern for their children was excessive screen time (Blandford, 2016).

The internet shows its ugly side when we see how easy it is for people to access pornography. In her article, "The lost boys—young men opt for virtual life of pornography and gaming," Cosima Marriner (2016) discusses the problem. She outlines how a generation of young men is being lost to arousal addiction, preferring the instant gratification offered by online pornography and gaming to the challenges of real-life interactions (Marriner, 2016).

This becomes even uglier when we look at violent pornography, where the effects can be even more damaging to young people. In an article "The evolving perception of human trafficking: your child's internet porn," Robert Benz (2016) interviewed Cordelia Anderson (Sensibilities Prevention Services, Minnesota, USA), where she stated:

> with Internet pornography being so accessible, it's the expectation that boys and men will be consumers—with girls and women as the objects to be consumed and to service men—otherwise their masculinity is challenged.

She goes on to say that (with the internet):

> there is, however, a substantial difference between the men's magazines of the past—Playboy, Penthouse and, even, Hustler—versus the dominantly violent content of Internet pornography with multiple sensory inputs and endless novelty.
>
> (Benz, 2016)

This quite clearly shows that it is "how" the technology is used or abused that gives it the good, bad or ugly tag.

Unmanned aircraft

Unmanned aircraft are no exception to this rule. They are also neat pieces of technology that every kid, old and young, would like to own. They are easily accessible to hundreds of thousands if not millions of people and, with their easy accessibility, governments, regulators and law enforcement agencies the world over have before them a challenge of epidemic proportion. For those that obtain the necessary approvals to use them and operate them safely, they are the good. For people who operate them illegally or too close to people and cause injuries, they are the bad. Now, enable the ability to operate such technology "anonymously" and you have what can only be described as a potential Mass Weapon of Destruction—the ugly.

When unmanned aircraft were first developed for military use they were used for missions that were considered "dull, dangerous and dirty." Now that they have found their way into the hands of everyday people, they are used to take aerial photographs of real estate and at sports events. They are used by hunters and anti-hunters. We see them being used by the police, fire fighters and the State Emergency Services (SES). The new group of "buzz words" for unmanned aircraft could very well be: Accessibility, Affordability, Adaptability, Anonymity and Autonomy. Both individually and collectively, these five defining characteristics are confronting governments and regulatory authorities throughout the world. By examining these attributes, we will be able to see more clearly the good, the bad and the ugly sides of this technology.

Accessibility

Small drones are more readily available, cheaper and easier to purchase than ever before, via internet stores such as Amazon and eBay as well as via all mainstream retailers. New manufacturers are popping up everywhere, locally, regionally and globally, often specializing in niche areas. FPV, multirotor or fixed wing options are a plenty. All major manufacturers, such as DJI Phantom, Parrot and 3DR are all selling online and via major retailers. It is also simple to purchase parts online and build your own for a very small cost. So, why are these devices more accessible and what is the effect on society to have greater accessibility to this technology?

Battery technology

The answer to the first part is the huge developments in battery technology. It is interesting to look at the history of batteries and see how far technology has come. While there are still some differing opinions, many believe that the first recorded use of a battery was in the city of Baghdad over 2,000 years ago. Known as the "Baghdad Battery," it was discovered among the relics found in a tomb near Baghdad in June 1938 by a German archeologist, Wilhelm Konig. The device consisted of a clay jar, which contained a copper cylinder that encased an iron rod. The jar was sealed with asphalt or pitch at its top opening. The height of the jar was about 13 cm, and the copper tube was about 12 cm in length and 4 cm in diameter. The vessel showed signs of corrosion, and early tests revealed that an acidic agent, such as vinegar or wine had been in the jar. Tests of replicas, when filled with vinegar, showed it could have produced between 1.5 and 2 volts between the iron and copper. It is suspected that this early battery may have been used to electroplate gold onto silver objects (BBC News, 2003). Rumpelstiltskin, eat your heart out.

There was a long wait for the next stage in battery development, which occurred in 1799 with the invention of the "voltaic pile" by Alessandro Volta (February 18, 1745–March 5, 1827). He published the details of his battery on March 20, 1800. His battery was made by piling up alternating layers of plates of silver or copper and zinc, separated by pieces of paper or cloth soaked in brine. When the top and bottom contacts were connected by a wire, an electric current flowed through the voltaic pile and the connecting wire. Interestingly, the French word for battery is "pile" (pronounced "peel").

There were many enhancements to Volta's original voltaic pile, with the Daniell Cell in 1836, the fuel cell, invented by William Robert Grove in 1839 with further improvements by Bunsen (who is better known for his burner), in 1942. The first rechargeable battery was the lead-acid battery invented by Gaston Plante in 1859. This type of battery is still used in cars today. The first commercially successful dry cell battery (zinc-carbon cell) was invented by Carl Gassner in 1881.

The battery that we are all probably most familiar with is the nickel–cadmium (NiCad) battery, which is a rechargeable battery that uses nickel oxide hydroxide and metallic cadmium as the two electrodes. The first type was invented by Swedish inventor Waldmar Jungner in 1899. The sealed NiCad type, which was developed around 1946, became the battery of choice for most of our power tools, portable electronic devices, flashlights and radio-controlled devices, such as model aircraft, throughout the 1990s as they could be manufactured in a wide range of sizes and capacities. NiCad batteries have a terminal voltage during discharge of around 1.2 volts, which remains almost constant until nearly the end of discharge thus providing the advantage of being able to deliver practically its full rated capacity at high discharge rates. Unfortunately, the "Cad" in NiCad does stand for Cadmium and while we were all enamored by how well these batteries worked in the early years, we eventually came to realize that it is a highly toxic metal and these batteries fell out of favor because of the environmental impact as a result of their disposal. Jungner probably realized the same thing and did run some experiments substituting iron for the cadmium, but he found that the nickel–iron combination was a poor substitute for the NiCad. Since Jungner was relatively unknown in the US, Thomas Edison patented his own NiCad battery in 1902, thus proving that he was never averse to "borrowing" other inventors' work. He also introduced the nickel–iron battery to the USA two years after Jungner had built one.

Model aircraft require larger currents of up to 100 amps or so to drive main motors. This was achieved from specially constructed NiCad batteries. It was possible to get about five or six minutes of model aircraft operation from quite small batteries, so a reasonably high power-to-weight ratio could be achieved. Interestingly, this power-to-weight ratio is comparable to an internal combustion engine.

The next big development in batteries came about with the lithium polymer battery (LiPo). The LiPo battery became the battery of choice for model aircraft due to their light weight. Their other big advantage is that they can be made into just about any shape. As advertised on the Padre Electronics Company website:

(The) Lithium polymer battery is a very powerful rechargeable lithium battery, safe material and packing, flexible shapes, light weight, small size and high energy density, no pollution, now has become the hottest battery for most digital and electronics use.

Padre's new generation LiPo batteries are as thin as 0.5mm and are available in a multitude of shapes. In 2009, 10.7 percent of mobile phones used LiPo batteries. By 2012, the ratio increased to 32.5 percent. In 2016 it is now 45 percent, and it is expected to be more than 60 percent by 2017 and many believe that they will eventually dominate the whole digital battery industry. LiPo batteries have a higher voltage per cell than NiCad, typically 3.7 volts (compared with 1.2 volts as discussed earlier). By combining these cells in either parallel or series, the current or voltage can be increased respectively. For example, three LiPo batteries connected in series provide a nominal 11.1 volts, which is a typical battery voltage found in many small drones.

Unfortunately LiPo batteries do have a bad side. They are a fire risk and are considered to be dangerous goods by the FAA, which issued FAA Notice N 8900.344, "Transportation of Lithium Ion and Lithium Ion Polymer Batteries as Cargo" on February 9, 2016. Dangerous goods are items that, when transported by air, can put people's health, safety or property at risk. They range from the obvious, such as explosives, radioactive items, and dangerous or volatile chemicals such as petrol and strong acids, to seemingly innocuous everyday items that can cause problems if not handled correctly, such as aerosols and lithium batteries. An incident occurred in 2014, when lithium batteries, which had been packed into a passenger's checked bag, short-circuited and started a fire in the aircraft's cargo hold. Fortunately, this occurred before passengers boarded the flight from Melbourne to Fiji. The passenger was a certified drone operator and should have known better.

The ICAO ANC recommended that certain lithium ion batteries be prohibited, (United Nations (UN) identification number 3480, Packing Instruction 965 only) on an interim basis as cargo on passenger aircraft, obviously until further studies could be conducted. The ICAO Council has confirmed this recommendation and the prohibition became effective from April 1, 2016. On the strength of this, CASA in complying with ICAO has new regulations, also effective April 1, 2016, that state:

- Batteries must be offered for transport at a state of charge not to exceed 30 percent of their rated design capacity. This means that you need to deplete your batteries before you attempt to transport them.
- A shipper is not permitted to present for transport more than one package in any single consignment.
- ". . . it is very important that LiPo batteries MUST be carried with you on board the aircraft and CANNOT be placed in checked in luggage! Swift changes in temperature and air pressure make LiPo batteries susceptible to catching fire. You don't want to end up like the owner of the carry case that caught fire on the Fiji Airways flight" (Dronethusiast, 2016).

The closing word on the latest in batteries on offer is the lithium–metal–polymer battery with ultra-thin film electrolyte. This lithium–metal–polymer battery has been developed for a number of applications including electric vehicle propulsion systems and batteries to power telecommunications installations. Because there is no liquid or paste electrolyte, they are maintenance free. They have service lives as long as ten years, under ambient temperatures from a range of -40°C to +65°C. As announced by Industrial Power Source (IPS) Batteries of Sante Fe, USA, a new thin-film construction for lithium–metal–polymer batteries allows the batteries to be embedded in printed circuit boards or integrated into circuit chips. The batteries are postage-stamp size and about 1/10mm thick, and contain "micro-energy cells." We have certainly come a long way since the "Baghdad Battery."

Effect on society

The second issue of greater accessibility to drones is the effect on society. Again, this can have its good, bad and ugly sides.

"It's important to open the skies to drones, and welcome them, but with an open eye towards the safety, national security and privacy issues," said John Villasenor, senior fellow in Governance Studies and professor at UCLA.

(Villasenor, 2012)

Only safety issues will be addressed in this chapter. The impact on privacy, security and from terrorism is covered in later chapters.

Let us first look at the numbers of drones out there flying in the skies over our heads. It is extremely difficult to put a figure on the number of drones sold around the world as there is not a lot of data collected, so any numbers provided here are really an educated guess, but it is interesting to bandy a few numbers about just for the fun of it. First off, we should issue a disclaimer; these numbers were pulled from sources and quotes all over the internet. There is a lot of "rumor in confidence," gossip and guesstimates, so all the conclusions are unverified and pure conjecture.

In the USA, Parrot (the drone manufacturer, that is, not the type you would find in your backyard that eats seed) made approximately US$53 million from drone sales in 2013, according to various websites. Assuming the average price of one of their drones is approximately US$300, this equates to around 180,000 units sold in 2013. Parrot's revenue was reported to have tripled in 2014, so an estimate of the number of units sold in 2014 is around half a million. If we assume that Parrot's revenue doubled again in 2015 that would equate to around one million units. We can probably assume that DJI and 3D Robotics, which are the other two big manufacturers, had sales in the same ballpark. Then, if we add in the smaller companies that are only selling thousands rather than tens of thousands, a reasonable estimate is that around four to five million drones were in the hands of everyday Americans in 2015 (Amato, 2015).

"There's a sense of excitement about this innovative new technology and what it can do," says Doug Johnson, Vice President of Technology Policy at the Consumer Technology Association (CTA), an industry trade group. The CTA estimated that 400,000 consumer drones would be sold during the Christmas 2015 season. This number may be conservative according to FAA official Rich Swayze, who told ATW Online in September 2015 that as many as one million drones could be sold during the holiday season (Karp, 2015).

The Teal Group is a respected corporation that conducts research and publishes accurate information on the aerospace and defense industry. The Group's market profile, released in August 2015, estimated that unmanned aircraft production will soar from the current worldwide production of US$4 billion annually to US$14 billion, totaling US$93 billion in the next ten years. Military UAV research spending would add another US$30 billion over the decade. According to Philip Finnegan (2015), Teal Group's Director of Corporate Analysis and an author of the Teal Group's study:

> Our coverage of the civil UAV market continues to grow with each annual report, mirroring the increase in the civil market itself. Our 2015 UAV study calculates the UAV market at 72 percent military, 23 percent consumer, five percent civil cumulative for the decade. Of the three areas, civil UAVs grow most rapidly over the forecast period as airspace around the world is opened.
>
> (Finnegan, 2015)

Again, it is difficult to put a number on drones sold in Australia in 2015, but since Australia has about 10 percent of the population of the USA, we can guesstimate that sales in Australia were around 10 percent of the USA, meaning that around 500,000 drones were purchased by Australian consumers in 2015.

Some of the good that is coming out of the greater accessibility of drones is the creation of jobs. The AUVSI in the USA estimates that 70,000 jobs will be created in the next three years alone, with the number increasing to 100,000 by 2025 (Toscano, 2013). Notwithstanding the positive comments about the growth of the industry, the numbers of drones coming into the airspace is a terrifying thought for regulators. It is frightening to imagine how tens or even hundreds of thousands of drones are going to be integrated into the airspace with manned aircraft. We know that drones can be used for purposes both good and bad. While most are used for sport and recreation on weekends by mum, dad and the kids flying around parks, many are also used for commercial purposes, both legally and illegally. In March 2016, CASA had issued 459 commercial operating certificates and anticipated that there would be in excess of 600 by the end of June 2016 (Skidmore, 2016). It would also be reasonable to guesstimate that there are at least the same number of operators who are not certified.

People flying model aircraft do not have to get approval to do so from CASA, nor do they need to join an organization such as MAAA. The MAAA has over 10,000 members, so again, a reasonable educated guess is that there are probably three or four times the number of people operating model aircraft who are not members of a dedicated club. In the USA, AMA has more than 140,000 members.

Whether people are operating legally, either for fun or commercially, or illegally, there is always the potential for accidents to happen. The most recent accident occurred on April 18, 2016 when British Airways Flight BA727 from Geneva was approaching Heathrow in the afternoon when what the pilot believed to be a drone struck the front of the aircraft. London Metropolitan Police said: "Thankfully the aircraft landed safely but the incident highlights the very real dangers of reckless, negligent and sometimes malicious use of drones."

Chief Superintendent Martin Hendy, Head of the Metropolitan Police Service's Aviation Policing Command said in a statement: "We continue to work with the Civil Aviation Authority and other partners to tackle this issue and ensure that enthusiasts who fly drones understand the dangers and the law."

In January 2015, a small drone crashed into a tree on the South Lawn of The White House grounds in the USA. In this case, the person was flying the drone for recreational purposes and lost control of it. In April 2015, a drone carrying traces of a radioactive material was found on the rooftop of the Prime Minister of Japan, Shinzo Abe's, office. On February 14, 2016, a drone crashed in the middle of a ceremony at the Australian War Memorial in Canberra. It was being flown during the evening Last Post ceremony. Fortunately, the person was identified and faces charges and a possible fine for breaching

CASA regulations (ABC News, 2016). Also in Australia, during a triathlon in Geraldton, Western Australia in 2014, a drone crashed into a triathlete who was competing in the race and she sustained head injuries. The drone operator was found to be in breach of CASA regulations and was fined AUS$1,700 for flying the drone within 30m of people (Taillier, 2014).

In August 2015 a drone was used to drop drugs into the prison yard at the Mansfield Correctional Institution in Ohio. The use of drones to smuggle weapons, mobile phones and drugs into prisons appears to be on the increase. This illegal operation in Ohio is not the only incident and similar cases have been recorded in Brazil, Ireland, Greece, Russia, Switzerland and Australia. The case in Australia occurred at the Metropolitan Remand Centre in Ravenhall, 22km west of Melbourne in March 2014, where the drone was carrying drugs to be dropped into the prison yard. Fortunately, police intercepted the drone before it could deposit its load (Caldwell, 2014).

Affordability

Technology in general gets more and more affordable each year. I remember my first flat screen TV, an LG 42-inch (106 cm) LCD TV that cost me AUS$3500 and weighed a massive 42kg; it took two people to carry it. Moving on a few years, my next purchase was an LG 60-inch (152 cm) LED Smart TV, which cost me AUS$2100 and weighs 23kg. I was able to mount it on the wall on my own. The new one has more features, a much better picture, I can watch 3D movies and I can connect to the internet. Personal technology such as mobile phones, computers and tablets are getting smarter, lighter, thinner, bigger and cheaper every year.

Drones are no exception. Amazon offers a range of high-technology drones equipped with camera for less than US$100.00. Aldi has also gotten into the game and was selling a small drone equipped with a camera in its stores at Christmas 2015 for AUS$99.00. An advertisement appearing regularly on Facebook shows a palm-sized drone with a camera for AUS$30.00. Walmart currently offers 19 drones on its website, the cheapest one going for just AUS$19.99.

As I sit here scratching my head, I can't think of anything bad or ugly about the price of technology going down, other than, I just wonder when the bubble is going to burst and we all wish for the good old days.

Adaptability

What do we mean by adaptability when we talk about the use of drones in society? Some synonyms for adaptable that come to mind in relation to drones are, flexible, versatile, alterable and multipurpose. Drones can be all of these and more. A small drone can be used on weekends for the kids to have a bit of fun with, and then on Monday, mum or dad can take the same machine and use it in the business to take photographs of real estate or to monitor developments on a construction site. The same unit can be used by the police, fire fighters, SES for search and rescue, scientific data collection, agriculture and farming, and monitoring the weather.

The drone is actually nothing special when it comes to any of the commercial activities. Manned aircraft and helicopters have been doing these jobs for years. It is all about the payload, i.e. the sensors onboard and the cost. Drones, depending on the size, can carry cameras, infrared/thermal imaging devices, gas sniffers, air quality sampling devices and the list goes on. The task at hand can then be accomplished generally at a fraction of the cost.

One of the new developments in the use of drones is the delivery of mail and small parcels to the door. This is a bit difficult to achieve with manned aviation, people would get a bit scared if a Robinson R22 tried to land in your front garden. But, it is very possible with small drones. Amazon and Google have been talking about it and carrying out some research over the past couple of years. Australia Post has started testing how it can deliver small parcels to its online shopping customers by drone. The trials are being conducted on Australia Post property in partnership with UAV manufacturer, ARI Labs (Coyne, 2016). Aside from your everyday customers, Australia Post is also targeting a few more specific situations in which drone delivery could really make a difference. In rural areas, for example, properties can often be a number of kilometers down a driveway from the actual road—and a drone could fly the package right down to the front door. Otherwise, they suggest it could be used for delivery of medication or other time–critical packages (Williams, 2016).

Anonymity

Drones are a bit like those stealth aircraft that are designed to avoid detection using a variety of stealth technologies that make them nearly invisible to radar, infrared or other detection systems. However, unlike aircraft like the F–117 Nighthawk, the B–2 Spirit or F–22 Raptor, drones by their very nature of being small and quiet just fade into the background. It is a real challenge to be able to spot a small quadcopter hovering at 400ft above your head. They are pretty inconspicuous. Then, we add the operator, who could be anywhere within a half kilometer radius and, to the average person in the vicinity, they are invisible to the naked eye. If we then add in the extra dimension of command and control of the drone via a satellite data–link, the operator could realistically be anywhere on the globe. This poses a huge problem to regulators who are trying their best to track drone usage and drone operators.

The FAA, CASA and other regulators around the world are introducing registration systems in an attempt to keep an eye on drone operators and drone operations. The FAA requires anyone who owns a small drone that weighs between 250g (0.55lbs) and 25kg (55lbs) to register with the FAA's UAS registry before they fly outdoors. People who do not register could face civil and criminal penalties. In relation to CASA's revised CASR Part 101, in force from September 2016, for persons intending to operate a drone commercially with a weight of 2kg or less in the standard operating conditions, he or she does not require either a UOC or a UAV Controller's Certificate. However, any person who is intending to operate a small drone must notify CASA of their operation. The regulation makes it an offense for a person to operate a very small drone for hire or reward without notifying CASA. The registration system will also allow CASA to establish and maintain a database of information that relates to these notifications. This all sounds a little futile, and while it gives the appearance of adding to safety, that is questionable. As stated earlier, there are already numerous illegal operators in Australia who have neither a UAV Controller's Certificate nor a UOC and to date I would be surprised if CASA has caught one of them. CASA, in its defense, has prosecuted a number of operators for breaches of the regulations, but these have all been legal operators and they have been prosecuted for operating outside their UOC. As for the USA, you don't have to be a rocket scientist to determine that, in a country with a population of over 320 million and an estimated four or five million drones in operation, there aren't a few illegal operators. Yet, despite all the trawling through the internet, we cannot find a single violation.

Autonomy

This is without a doubt the least understood and the most misused term in the unmanned aircraft world. The main issue here is that people are confused between the terms "autonomy" and "automation." The *Collins Dictionary* defines autonomy as, "freedom to determine one's own actions, behaviour, etc." and further defines autonomous as, "independent of others." Dictionary.com gives a better definition for autonomous, "having autonomy; not subject to control from outside; independent." Automation, on the other hand, means "the technique, method, or system of operating or controlling a process by highly automatic means, as by electronic devices, reducing human intervention to a minimum." Thus, autonomy means that the human is replaced; automation does not. ICAO defines autonomous aircraft as "an unmanned aircraft that does not allow pilot intervention in the management of the flight" (ICAO Doc 10019, 2015) If we consider that an autonomous aircraft does not allow pilot intervention, then it is uncontrolled. However, an automated aircraft does, at certain phases of its flight, receive inputs from a remote pilot. The aircraft is then allowed to fly some pre-planned maneuvers before receiving further commands. Automation can assist in reducing the amount of human intervention required and can improve the quality, accuracy and precision of a machine. In relation to drones, the ability to automate some aspects of the operation can result in a safer overall operation.

Autonomy requires very sophisticated artificial intelligence that will allow the drone to be capable of intelligent behavior. To achieve this, the unmanned aircraft will have to perceive its environment, make the necessary decisions and take the appropriate actions without any human involvement. To achieve autonomy, some of the attributes that an unmanned aircraft will need include the ability to think, reason and learn, and we are a long way from this. An article in the *Air Force Magazine*, takes a more realistic view:

> Remotely piloted aircraft such as the MQ-9 Reaper and RQ-4 Global Hawk are manned by squadrons of pilots and sensor operators on the ground. Five or 10 years from now, however, that may no longer be the case, as full autonomy for air vehicles is well within the Air Force's technical reach. According to USAF officials, artificial intelligence and other technology advances will enable unmanned systems to make and execute complex decisions required for full autonomy sometime in the decade after 2015.
>
> (Grant, 2014)

Mission flight planning is in effect automating or replicating the mission flight planning capabilities of a human pilot at the strategic and tactical levels. The ability to automate this aspect of the operation has the potential to substantially increase the level of "autonomy" onboard the unmanned aircraft, resulting in safer overall operation. Consider different levels of automatic decision and action selection as described in Table 5.1 (Parasuraman *et al.*, 2000). The table has been reversed from the original document.

In the real-time environment of drone operations, it is more effective to be able to operate at the sixth level in this model; that is, the entire mission (from take-off to landing) is executed automatically unless there is human intervention. This is because unmanned aircraft generally operate in a highly dynamic environment far removed from the human operator who is on the ground or in another aircraft. The remote pilot could even be in another country. Changes in the environment may necessitate a change to the current flight plan but any changes must be made within a limited time interval. An unmanned aircraft operating at the fifth level of autonomy could potentially miss out on an important window of opportunity (due to unreliable, low-bandwidth communications links typically found in UAS) if the human operator cannot respond in time. However, UAS operating at the sixth level can

Table 5.1 Levels of automation of decision and action selection

1	Computer offers no assistance; human makes all decisions and takes all actions
2	Computer offers a complete set of decision/action alternatives
3	Computer narrows the selection to a set of alternatives
4	Computer suggests one alternative
5	Computer executes suggestion with human approval
6	Computer allows the human a restricted time to veto before automatic execution
7	Computer executes automatically, and then informs the human
8	Computer informs the human only if asked
9	Computer informs the human only if it decides to
10	Computer decides everything, acts autonomously, ignores the human

respond immediately if no human operator input is received. Additionally, a UAS operating at level six has a higher degree of immunity from failures in communications as the mission can still continue without human operator input. This occurs as actions are executed automatically unless a human operator intervenes. Military unmanned aircraft such as the MQ-9 Reaper and the RQ-4 Global Hawk are probably operating at the sixth level. If we ever get an aircraft operating at level ten, we could get a situation where the pilot on the ground says, "Taking Over" and the drone responds with, "No thanks, I've got this."

Although the emphasis has been on automation, it is recognized that there are increasing degrees of autonomy in unmanned aircraft, however, no formal definitions of scale or recognized methods of certification exists from any regulatory authority. So, before we can have full autonomy, a number of things have to happen; artificial intelligence has to be developed to a much higher level than currently exists, manufacturers will have to design and build the computers or the robots, or the drones that are capable of intelligent behavior, and regulators will have to develop the standards and regulations to manage the process.

References

ABC News "Drone Crashes in Middle of Ceremony at Australian War Memorial in Canberra, February 19, 2016, www.abc.net.au/news/2016-02-19/drone-crashes-in-middle-of-ceremony-at-war-memorial-in-canberra/7185202, accessed August 16, 2016.

Amato, Andrew "Drone Sales Numbers: Nobody Knows, So We Venture a Guess" 16 April, 2015.

BBC News "Riddle of Baghdad's Batteries" February 27, 2003.

Benz, Robert J "The Evolving Perception of Human Trafficking: Your Child's Internet Porn" The *Huffington Post*, April 12, 2016, www.huffingtonpost.com/robert-j-benz/the-evolving-perception-o_4_b_9627416.html, accessed August 15, 2016.

Blandford, Megan SBS, February 10, 2016.

Caldwell, Alison ABC News, March 10, 2014.

Coyne, Allie "Australia Post Trials Parcel Delivery by Drone" April 15, 2016.

Dronethusiast. www.dronethusiast.com, updated March 2016, accessed May 15, 2016.

Finnegan, Philip "Teal Group releases 2015 UAV Market Profile and Forecast" Teal Group Press Release, August 17, 2015.

Grant, Rebecca *Air Force Magazine*. April 2014.

International Civil Aviation Authority (ICAO) Manual on Remotely Piloted Aircraft Systems (RPAS) (Doc 10019)1st Edition 2015, www4.icao.int/demo/pdf/rpas/10019_cons_en%20-%20Secured.pdf, accessed August 15, 2016.

Karp, Aaron "FAA Warns of a Million Drones under People's Christmas Trees" September 28, 2015.

Kaushik, Preetam "Drones in the USA: The Battle for the Civilian Market" September 20, 2013.

Marriner, Cosima *Sun-Herald*, April 17, 2016.

MGM Studios, Inc. v *Grokster, Ltd.*, 545 US 913. 2005, www.law.cornell.edu/supct/html/04-480.ZS.html, accessed August 16, 2016.

Parasuraman Raja, Sheridan, Thomas B. and Wickens, Christopher D "A Model for Types and Levels of Human Interaction with Automation".

Seo, Choi Kyung SBS, 2016.

Skidmore, Mark CEO CASA. AAUS Keynote Address, Canberra, March 7, 2016.

Taillier, Sarah ABC News, November 13, 2014.

Toscano, Michael AUVSI CEO, March 12, 2013.

Villasenor, John Senior fellow in Governance Studies and Professor at UCLA, as quoted in "The Impact of Domestic Drones on Privacy, Safety and National Security", April 4, 2012, www.brookings.edu/events/the-impact-of-domestic-drones-on-privacy-safety-and-national-security/, accessed August 16, 2016.

Williams, Hayley "Australia Post Has Completed First Trials Of Drone Delivery Services" 15 April 2016.

Woo, Chae Chan SBS, 2016.

6 Eyes in the sky
Invasion of privacy

> Privacy is not something that I'm merely entitled to, it's an absolute prerequisite.
> Marlon Brando

Protection of an individual's privacy is undoubtedly the most contentious issue facing civil drone usage throughout the world. Drones have the potential to pose a serious threat to a person's privacy or a business's commercial activities. As drones become cheaper and more capable they can accommodate high-resolution, lightweight digital cameras with audio recording devices. This new mode of surveillance provides a cost-effective way for governments, companies and individuals to observe and collect information on citizens—potentially without their knowledge or consent. In the US privacy issues were the primary reason for delays to a number of the government's mandated deadlines to integrate drones into the NAS.

Invasion of privacy is just one of many areas in which existing laws are challenged but it's this area that the chapter will concentrate on. The reason for such a focus is because most invasion of privacy incidents are those operating in populous areas, at low levels and with small or micro-drones. And this is the precise airspace in which those regulators who have regulated have permitted commercial and recreational drone activities.

The invasion of privacy can be either "intentional," as in the case of surveillance or "Peeping Toms," or "inadvertent" such as with activities such as aerial photography (including Google Earth), crowd and traffic monitoring, or with search and rescue missions. For instance, the USAF policy permits the "incidental" capture of domestic imagery by drones. However, the policy would not ordinarily allow the targeted surveillance of a US citizen unless explicitly approved by the Secretary of Defense consistent with domestic law and regulations.

One of the main factors that have led to the rapid escalation of privacy concerns is the fact that drones are far stealthier and more sophisticated than other means of surveillance. They can accomplish such tasks much more effectively than can, for example, helicopters. The use of drones for surveillance in the US, Europe and Asia has been the most common and prolific application—and mostly well before they have been permitted to be operated commercially. With purpose-designed equipment, drones can intercept communications, peer through windows and can carry facial recognition technology. The potential for violations of civil liberties is indeed most significant and a very real threat.

Each country that has drone regulations has different rules relating to their operations and what restrictions may apply. One area of aviation law that does have a high degree of harmonization concerns the lower limit of what is generally termed "navigable airspace." In the US, Congress has defined navigable airspace as "airspace above the minimum safe altitudes of flight prescribed by the Civil Aeronautics Authority" and further provided that "such navigable airspace shall be subject to a public right of freedom of interstate and foreign air navigation" (Air Commerce Act 1926 as amended by the Civil Aeronautics Act 1938).

Essentially navigable airspace is that airspace in which aircraft operate with the lower limit usually prescribed at 500ft AGL. Most aviation regulators—or at least those that have promulgated RPAS regulations—allow for a nominal 100-ft buffer zone in permitting drone operations up to 400ft AGL. Obviously they also impose restriction on drone usage within the vicinity of aerodromes or heliports or in controlled or restricted airspace.

Many aviation regulators have taken the approach that the UAS community can play a critical role in educating the broader public and engaging in meaningful dialogue with them to demonstrate the positive aspects of UAS technology and the benefits that can be provided to society. It remains to be seen how aviation safety regulators like the US FAA and the EASA will effectively control drone activities without duly considering the impact of privacy concerns of their operations, especially given the delays being experienced by the FAA to achieve its government's corresponding mandate have arisen primarily due to privacy concerns.

Jurisdictional issues

One of the major challenges for governments with respect to privacy issues relates to the jurisdiction of aviation regulatory authorities. Dealing with privacy matters is usually not part of the statutory role or function of aviation safety regulators throughout the world. The protection of privacy in most countries is governed on a national or state level by specific privacy legislation. Moreover, most existing privacy regulations govern how a government agency or commercial enterprise is to collect, store, use, disseminate and protect a citizen's personal information. Most often such legislation does not extend to protect citizens from private individuals who collect personal information (including video images) using unmanned aircraft.

At the local or community level, municipal councils or local governments have jurisdiction over public land, but do not have jurisdiction over the airspace above it. And while the aviation regulators can regulate safety, they usually do not have jurisdiction over privacy matters. Herein lies the dilemma.

The key consideration in respect of privacy protection is to protect citizens' rights to privacy by reviewing and aligning relevant legislation. Nominating the interrelationship that must exist between safety regulators, privacy agencies and other key stakeholders would facilitate that review and alignment, and the subsequent development of UAS aviation safety regulations.

While ICAO provides a forum to coordinate air safety issues, member states must consider domestic implications of the use of drones, including the privacy of its citizens. Generally, the international norm for privacy flows from Article 12 of the Universal Declaration of Human Rights 1948, which provides that "no one shall be subjected to arbitrary interference with his privacy." The issue that arises is whether UAS are permitted to operate in national airspace over public places (such as parks and beaches) and, if so, will there be a requirement for UAS to avert its "eyes" from the ground?

In the US, commentators have argued that interference with privacy would contravene the US Constitution's Fourth Amendment, in that it protects US citizens against unreasonable searches and seizures. In this context, the FAA held public consultations as far back as 2013 to seek comment on draft privacy provisions for its UAS test site. While the FAA does not intend for the privacy provisions to apply more generally across non-segregated airspace, it may assist future discussions on how law, policy and industry practice should respond in the longer term.

All countries struggle with privacy issues arising from drone operations. In Australia, which was the first country to develop UAS regulations, the problem persists. The Australian Privacy Commissioner recently commented that the Privacy Act 1988 (Cth) does not contain provisions dealing with invasion of privacy from individuals operating UAS. Australia's aviation safety regulator, CASA, similarly acknowledges the regulatory gap between CASA's focus on air safety and the wider Commonwealth's legislative responsibility to ensure and protect the privacy of its citizens. As was highlighted earlier in this chapter, the area of most concern is in situations where the operation of the drone is below navigable airspace.

As we will consider later in this chapter, in the English-speaking world only the courts of the USA have developed a tort of invasion of privacy. Notably, the Australian Law Reform Commission in reviewing the US's experience has undertaken an inquiry into "Serious Invasions of Privacy in the Digital Era" and reviewed, among other things, the growth in capabilities to use surveillance and communication technologies and community perceptions of privacy. As privacy considerations through the use of drones gather momentum in jurisdictions throughout the world, hopefully this will generate a sufficient impetus to review the adequacy of privacy legislation in the new age of aircraft.

A drone perspective

Viewing our world from the "eyes" of a drone can provide an entirely different perspective. Even everyday activities and features when viewed from a different angle take on an entirely new outlook. For example, if you were to sit in a wheelchair and go through your home you would see an entirely different perspective of your house. Also see how different your house looks from a drone perspective. The main reason why things appear so different is directly related to how we learn and how our brain processes information. Although this topic is discussed in Chapter 8 on human factors, it is worth briefly considering the psychological processes involved to better understand the reason why drones are having such a profound effect upon the way we see the world.

Cognition and cognitive process has long been the domain of psychological studies and research. Cognition is the mental action or process of acquiring knowledge and understanding through experience, thought and the senses. Perception on the other hand is the ability to become aware of something through our five senses of seeing, hearing, touching, tasting and smelling. Perceptive processes describe the way in which people translate sensory impressions into a coherent and unified view of the world around them.

Psychologists have found that approximately 80 percent of what we learn is acquired through our sense of sight. What we hear accounts for, on average, 13 percent. Moreover when we perceive things through multiple senses the amount we learn increases significantly. This is the reason why audio visual advertising, such as television commercials, are far more effective than collectively seeing an advertisement in the newspaper and hearing the same information on the radio. Considering the ability of small drones to be equipped with high-resolution cameras and audio detection devices—the user has a very powerful device for acquiring new and previously unattainable information.

Twentieth-century British novelist C.S. Lewis could never have imagined the insightfulness of his comment when he stated: "What you see and what you hear depends a great deal on where you are standing. It also depends on what sort of person you are."

The expectation of the law in respect to drones is that there is a human observer and a human being observed, and the observation made is deemed by the courts to be with the naked eye. So if drones can provide us with an entirely different standpoint then we are well positioned to learn new and different things from what we "see"—even about that with which we are already familiar. But where this capability of drones becomes a problem to society is when the information attained is of a private nature and possibly attained without the individual's knowledge or consent. Certainly Mr Lewis did not have in mind the prying eyes of an adolescent neighbor drone operator when he penned these insightful words.

Common law protection of privacy

The law of torts has developed in the courts of countries that have adopted the British common law system. Tortious cause of actions impose civil liability on persons by way of compensating, in certain instances, the innocent parties who have, in some way, suffered injury as a consequence of another person's wrongdoing.

Tort, a French word, literally means a "wrong." Within a legal context it refers to a civil wrong. The plaintiff of a tortious cause of action will claim that the defendant (the tortfeasor or wrongdoer) has in some way violated or interfered with that person's rights. Unlike a criminal action, the wrong alleged to have been committed is not against the state—although it is deemed to be socially irresponsible—but is a wrong against the individual.

The torts of trespass to land and private nuisance correlate insofar as they both pertain to interference with an occupier's exclusive use of land. Trespass to land relates to direct interference with that exclusivity while private nuisance relates to indirect interference. Trespass to land includes the entry on to land (or the airspace above it) without permission, remaining after permission to stay has expired and leaving things on the land without authority to do so.

Traditionally, in respect to vertical limits of one's land ownership, customary law had applied and was based on an ancient Roman maxim *cujus est solum ejus est usque ad coelum* meaning: "Whose is the soil, his is also that which is up to the sky." Likewise this legal principle had long recognized the absolute sovereignty of the state over its territorial airspace extending to an unlimited height. This universal right of national sovereignty has also been codified in over 190 states in their ratifying of the provisions of, initially, the Paris Convention in 1919, then later the Chicago Convention of 1944 both of which relevantly provide in Article 1:

> The High Contracting Parties recognise that every Power has complete and exclusive sovereignty over the air space above its territory.

In the US, almost at the dawn of commercial aviation, Congress enacted the Air Commerce Act of 1926, confirming that the USA has "to the exclusion of all foreign nations, complete sovereignty of the airspace" over the country while citizens had "a public right of freedom of transit in air commerce through the navigable air space of the United States."

In respect to an individual's property rights, and again from the earliest (hot-air balloon) days of aviation, conventional legal principles were challenged by the unique capabilities of aircraft. The traditional notion was that ownership of private property included all of the soil beneath, and all of the airspace above one's land. This was inconsistent with the aspirations of aviators who required free passage across private property boundaries to enjoy the technological benefits conferred by point-to-point aerial navigation.

Long before modern commercial air transportation and air traffic control regulations and practices existed, it was left to the courts to deal with manned aircraft flights by reference to the legal concept of trespass. In arguably the first aviation decision in this area of the law, the case of *Guille* v. *Swan* involved a property owner in the early 1800s. Guille brought an action after the operator of a hot-air balloon crash-landed into his garden in New York City. When the balloon descended the balloonist called for assistance and more than 200 persons broke into the garden through the fences, and came onto the premises; beating down vegetables and flowers. The court found the balloonist strictly (no fault) liable for trespass and the property owner recovered an award of money in damages.

One common law concept that has come increasingly under challenge since the advent of the aircraft is encapsulated in the saying "a man's home is his castle." This legal principle derives from English common law in protecting one's home against external intrusion. The parallel legal doctrine in the USA is known as "Castle law," or "Defense of habitation law." This principle of law recognizes one's residence as a place within which one can enjoy and reasonably expect protection from illegal trespass and violent attack.

By as early as 1936, the US courts in the case *Hinman* v. *Pacific Air Transport* found that this principle was not unlimited—such a notion was not the law and never had been. The court held, *inter alia*:

> We own so much of the space above the ground as we can occupy or make use of, in connection with the enjoyment of our land. This right is not fixed. It varies with our varying needs and is coextensive with them. The owner of land owns as much of the space above him as he uses, but only so long as he uses it. All that lies beyond belongs to the world.

In the UK in 1978 there was a landmark decision with the case of *Bernstein of Leigh* v. *Skyviews*. In this case the defendant was an aircraft charter company that took a single aerial photograph from a light aircraft while overflying the plaintiff's country residence. In bringing an action for trespass the plaintiff, Baron Bernstein of Leigh, claimed, as owner of the land and the airspace above it, that he had a right to exclude any entry into that airspace. The court dismissed the case and held that the landowner's rights did not extend to an unlimited height. The judgment stated:

> The rights of an owner in the airspace above his land is restricted to such height as is necessary for the *ordinary use and enjoyment of his land and the structures upon it* (Emphasis added).

This case has become the leading precedent, in most common law countries, regarding trespass to land claims. Its application to drones is that no action will arise against the operator or controller of an unmanned aircraft overflying an occupier's land at normal, prudent cruising levels. Therefore, the ancient definition of land as extending from the ground to the extremities of the heavens has since been modified by the courts and, as we shall see in the following sections, by Parliament.

Legislation in all countries provides for overflight of aircraft over all property. In respect of horizontal or lateral limits of sovereignty, international treaties—

which have subsequently been ratified by Parliaments into domestic law—have clarified the situation. Article 2 of the Chicago Convention 1944 states:

> For the purposes of this Convention the territory of a State shall be deemed to be the land areas and the territorial waters adjacent thereto under the sovereignty, suzerainty, protection or mandate of such State.

With respect to drone activities, a cause of action in trespass may arise against the operator of an unmanned aircraft that overflies private land at a sufficiently low level such that it interferes with the occupier's "ordinary use and enjoyment" of that land. This action may arise even if the drone operator is complying with exemptions for low-level flying. Similarly, private-nuisance claims may arise against the operator of a drone that flies over, or in proximity of, an occupier's land at lower-than-permissible levels, including when taking off and landing. Even at normal altitudes a private-nuisance action may arise when substantive damage exists. However, where the UAS operator can show that the act was done by a stranger or pursuant to statutory authority—such as, for example, police surveillance—the private-nuisance action is unlikely to succeed.

The key consideration in the torts of trespass of land and private nuisance relates to UAS operating from semi-prepared or *ad hoc* areas. This raises the question as to how an occupier of land levies the above tortious actions against UAS operators if those operators are not readily identifiable by their unmanned (and unmarked) aircraft.

The potential for anonymity of drone operators is a real issue for our legal system and for society at large. As an unmanned aircraft could be used by an operator to stalk or harass citizens, in the same way that would breach a citizen's right to privacy. It is therefore important that governments and the courts consider the provisions needed within the body of criminal law to accommodate these acts of intimidation when specifically performed using drones.

In the English-speaking world only the USA has developed a tort of "invasion of privacy" solely by judicial decision. The US courts started the development of this tort in the latter part of the nineteenth century, following the famous article "The Right to Privacy" by Warren and Brandeis (1890) that synthesized mainly English decisions on the torts of defamation, nuisance and trespass. In other countries, Parliaments have introduced various legislation in an attempt to protect the privacy of individuals. For example, the statutory tort of violation of privacy has been introduced in Canada. But once again these laws were introduced at a time when the activities of drones in society were beyond contemplation.

Overflight of satellites

As the resolution capabilities of cameras onboard satellites continue to increase so too do the concerns of individual's privacy violations on the ground. In Europe, satellite images and aerial photographs have been used to monitor farming activity. Google Earth's observation satellite cameras and sensors capture everything in their wake as they circle the globe. Deliberate or otherwise, all activities on the ground are captured on a regular and cyclic basis. Moreover, when this data is analyzed and combined with other spatial and thematic input, the resulting images and information assume new dimensions. It is also possible to deploy instruments and cameras with facial recognition capabilities.

Legislation in many countries has required Google to de-identify certain images that enabled individuals and their property (for example car registration markings) to be identified. International treaties have modified the position in regard to satellites in asserting that "no national appropriation by claim of sovereignty" can prevent overflight rights of satellites in outer space, notwith-standing that no precise definition of outer space is provided. Interestingly, this is one of the few areas of aviation law that has developed through customary international law.

An important aspect of customary international law is that it applies universally among nation states, regardless of whether such laws have been formally legislated or codified. It is thought that the flight of the first satellite—the Soviet Union's Sputnik I back in 1957—gave sufficient grounds for the establishment of a rule of customary international law allowing a space object of one state to overfly the territory of another state.

The launching of Sputnik I was a matter of notorious fact around the world, and especially in the international scientific community. The satellite itself was visible in the night sky as it flew overhead (if one knew where to look), and it was aloft for several months. During its mission, Sputnik I made hundreds of orbits, and overflew many countries a great number of times, but not one state made any official protest or reserved its right to do so.

The inaction by states satisfied the two elements required for the development of a rule of customary international law. The elements were, first, that the states that were overflown were aware of the overflight, and second, there were many opportunities for them to make complaint about any one of the overflights. Eminent scholars in the field have said that this was sufficient to create, almost instantly, a customary rule of international law allowing freedom of overflight in outer space.

National privacy laws

Various states throughout the world have introduced privacy acts and other related legislation long before the proliferation of drones into society. The laws generally apply to the activities of the respective governments and their agencies and tend to be limited to those entities. Some of these laws, though not necessarily by design, may protect against invasive or inappropriate use of drones. For example, many states have legislation that makes it illegal in certain circumstances to use a surveillance device to record or monitor private activities or conversations via listening devices, cameras, data surveillance devices or tracking devices.

Some countries have existing laws that can prohibit the use of drones where issues of privacy are concerned. For example, in the USA "Peeping Tom" laws make it illegal to use a camera-mounted drone to spy on a neighbor's backyard sunbathing habits. This is based on the premise that a neighbor has a reasonable expectation of privacy protection under the Fourth Amendment of the US Constitution and is justified in the belief that no one should be observing from above. If litigated the neighbor would most likely sue for "intrusion upon seclusion." This is considered to be highly offensive behavior to a reasonable person. Section 652B of the Restatement (Second) of Torts creates a cause of action for intrusion upon seclusion. Other examples of legal restrictions in the US on the use of drones are apparent from the court cases reported below.

The main problem, however, is that with most existing privacy laws, they simply are not appropriate for modern communication technology—let alone rampaging UAS technology. To better illustrate the type of problems commonly encountered with current privacy legislation, and how they are ineffective in dealing with issues relating to drone activities, the following case study is provided.

"Eyes in the sky" inquiry into drones and privacy

The Australian Council of Civil Liberties called on the Australian government to urgently deal with the privacy issues associated with drones. Regulations governing civilian drones had not kept pace with the rapid growth of the industry in Australia. This was probably largely attributable to the fact that Australia was the first county in the world that permitted the commercial civil use of drones when regulations were promulgated back in 2003. What has happened in Australia is therefore a good barometer as to what is likely to be experienced throughout the world as and when RPAS regulations are introduced.

In July 2014 the Australian Commonwealth Parliament's House of Representatives Standing Committee handed down its report entitled: "Eyes in the sky: Inquiry into drones and the regulation of air safety and privacy." The inquiry reviewed the emerging issues around drone use and examined the adequacy of the existing legal and regulatory framework with a particular focus on safety, privacy and security issues.

During a series of hearings and roundtables, the Committee heard from CASA about the importance of allowing drone technology to mature so that the risk to people and property is minimized. The Committee also heard from privacy experts, including the Commonwealth Privacy Commissioner, about the complexities and gaps in Australia's privacy laws and the inadequacy of law—both federal and State—to protect individuals against privacy invasion from drone activities. Essentially the Committee found that privacy laws are not sufficiently robust to ensure that the new technology is used responsibly and is consistent with democratic values.

In Australia the primary legislation for the control of privacy issues is the Commonwealth Privacy Act 1988. This statute, like so many similar laws throughout the world, provides a number of privacy protections to the public and applies to most government agencies and many private sector organizations. The 13 related "privacy principles" are very similar to those in the US and govern how organizations should collect, use, disclose, provide access to and secure personal information. However, the Act does not provide citizens with comprehensive privacy protection.

The Privacy Act, again like so many similar statutes throughout the world, does not apply to the collection and use of personal information by private citizens and does not provide overarching privacy protection for the individual. This is precisely the area where the use of small and micro-UAS can result in the serious invasion of an individual's privacy. However, the Act was never designed or intended to protect against such intrusions into citizens' private seclusion. Dr Roger Clark from the Australian Privacy Foundation told the inquiry:

> We identify privacy of personal behaviour . . . as the interest that people have in not being intruded upon by undue observation or interference with their activities, whether or not data is collected—after which it would then move into another space. When we look at the Privacy Act . . . it is all but irrelevant to behavioural privacy protection. It was designed that way; it was designed to deal with data protection only.

So although the Privacy Act offers substantial privacy protections in certain circumstances there are a number of situations in which it may not protect Australians against the invasive use of drones. The Committee concluded that issues arising from the expanding use of UAS would require significant changes to both federal and State privacy law.

The Australian Law Reform Commission's Professor Barbara McDonald agrees with the inquiry's findings in that the exemptions contained within the Act and the "patchwork" of State and federal privacy laws are totally inadequate to deal with drone operations:

> At the moment the lack of uniformity means that there is insufficient protection of people's privacy, because people do not know what is against the law and what is not. But it is also insufficient protection for organisations like those in the media.

The report made two recommendations in respect of privacy issues arising from the inquiry. The first suggested that CASA should provide more information to users and manufacturers regarding privacy issues. The second recommended that the Australian government consider introducing legislation by July 2015 to provide protection against privacy-invasive technologies and to consider the Australian Law Reform Commission's proposal for the creation of a tort of serious invasion of privacy.

In further developments the Australian Law Reform Commission has subsequently outlined a proposed remedy in a discussion paper on serious breaches of privacy in the digital era, which involves a new tort of privacy. If such a proposal was put into law, a person could sue for a serious invasion of privacy if their "seclusion or private affairs" were intruded upon, or if private information about them was misused or disclosed.

If the above recommendations and initiatives are implemented they could have far-reaching effects and would presumably go a long way toward addressing the privacy issues arising from the widespread, and increasing, use of drones in society. Paradoxically it has been suggested that while there is a perception that the extensive use of drones—especially for surveillance—may erode the privacy of citizens, the opposite may in fact occur. Drones could be just what the doctor ordered in providing the "visceral jolt" that society needs to drag privacy law into the twenty-first century and help reconceptualize the mental model of privacy violations.

Other legal issues

Invasion of privacy is not the only legal issue confronting society in respect of drone operations—it is just the most apparent and confronting issue. The issue of accident responsibility and liability involving UAS is another important area of the law and one that is not adequately addressed by either the ICAO guidance material or regulations developed by any national aviation authority.

Annex 13 of the Chicago Convention defines an accident as follows:

> An occurrence associated with the operation of an aircraft which takes place between the time any person boards the aircraft with the intention of flight until such time as all such persons have disembarked, in which . . . [*inter alia*] a person is fatally or seriously injured.

While a collision between a manned aircraft and a UAS would comply with the definition of an "accident," the collision of two UASs or the impact of an UAS with property or equipment would not be addressed under the current legal framework. Notably, the ICAO Technical Commission discussed this issue at the 36th Assembly, stating that, despite the best efforts of regulators, manufacturers and operators, accidents will occur.

The UK CAA literature on this issue indicates that while the strict definition of "accident" or "incident" may not apply to drones, the guidance material sets out the types of "reportable occurrence" that may apply specifically to UAS, including loss of control or data-link, navigation failures, structural damage and flight programming errors.

Increasing the complexity of apportioning liability, the question arises— would the operator or the manufacturer be at fault for UAS operating in autonomous mode? Moreover, if a UAS was operating autonomously on defective software, does this reduce a human operator's responsibility for failure to detect and avoid? The legal issue that arises is whether the causal link between the operator's action (or omission) and the resulting drone accident is sufficiently established to remove doubt about apportioning liability. Besides creating dilemmas for accident investigators and insurers, it will also be difficult for lawmakers to design a strict (no fault) liability regime—currently applied to manned aircraft accidents—for UAS-specific scenarios.

Addressing the issue of privacy protection

Although there are a myriad of other legal-related issues arising from the civilian operation of drones, this is not the specific focus of this chapter. Undoubtedly drones are able to undertake assignments that were previously impossible, and they can reduce the cost—and the risk—of many "dull, dirty or dangerous" missions. However, like any new technology, drones can be misused. As we saw in Chapter 3, drones can pose a safety risk to other aircraft or to people and property on the ground, but it is their increasingly sophisticated and invasive cameras and sensors onboard that can be used to compromise the privacy of both citizens and businesses.

Rapidly increasing drone availability and diversity of applications raises serious privacy issues for governments and their citizens. With the ever-increasing accessibility and affordability of drones both locally and via the internet, it will become more and more difficult to enforce regulatory compliance.

The challenge faced, not only by governments but for the community as a whole, is to realize the potential of this innovative technology while protecting against its risks—and in particular the privacy-invasive risks. The implementation of voluntary codes of conduct and privacy policies may go some way to address public concern about the potential for invasive drone use. For instance, in the US the Association for Unmanned Aircraft Systems International have developed a code of conduct that asks its members to respect individuals' privacy. However, it is more likely that it will be up to respective governments throughout the world to introduce "drone-specific" legislation to address the issue of privacy-invasive technologies.

A final word of warning

It is important to realize that when developing legislation, while the focus may be on drones, privacy concerns relating to new technologies is a far broader issue. In providing comment to the previously cited "Eyes in the sky" inquiry, Dr Reece Clothier argued that, instead of focusing on the privacy threats posed by drone activities, it is important for Parliaments to take a broader view of how privacy is affected by technological advances:

> We need to step away from this idea that it is a specific piece of technology or a specific device and say, "Let's protect the interests of privacy" . . .

Google Glass is a much more invasive technology that every person is going to be wearing in the next five years. So whether it is drones, Google Glass or the fact that I can collect metadata on your Facebook account and marry that up with your LinkedIn and actually track your movements, it is your personal information . . . it is an issue much broader than unmanned aircraft.

So the arrival of the drone age brings with it some unique social, ethical and legal challenges. These challenges highlight the inadequacies of current privacy and surveillance laws throughout the world. However, at the end of the day the biggest problem may not necessarily be with drones *per se*, but with the rampant development of privacy-invasive technologies. It is therefore imperative when developing new legislation to distinguish between those issues that are drone-specific and those that are not. To throw the baby out with the bathwater will mean that society, as a whole, will lose out on the benefits that only drones can bestow upon humanity.

References

Background Briefing "Drones Fly through Privacy Loophole" ABC News (online), September 14, 2012, www.abc.net.au/news/2012-09-13/drone-technology-prompts-privacy-law-review-call/4260526, accessed August 15, 2016.

Dolan, Alissa and Thompson, Richard *Integration of Drones into Domestic Airspace: Selected Legal Issues* (Report No R42940, US Congressional Research Service, 2013) 29, www.fas.org/sgp/crs/natsec/R42940.pdf, accessed August 15, 2016.

Griffith, Chris "Drones a Safety and Privacy Headache" *The Australian* (online), July 18, 2013, www.theaustralian.com.au/australian-it/personal-tech/drones-a-safety-and-privacy-headache/story, accessed August 15, 2016.

House of Representatives Standing Committee on Social Policy and Legal Affairs "Eyes in the Sky: Inquiry into Drones and the Regulation of Air Safety" The Parliament of the Commonwealth of Australia, Canberra, July 2014.

Kapnik, Benjamin "Unmanned but Accelerating: Navigating the Regulatory and Privacy Challenges of Introducing Unmanned Aircraft into the National Airspace System" 2012 77(Summer) *Journal of Air Law and Commerce* 439.

Roberts, Troy "On the Radar: Government Unmanned Aerial Vehicles and their Effect on Public Privacy Interests from Fourth Amendment Jurisprudence and Legislative Policy Perspectives" (2009) 49 *Jurimetrics* 491–518.

Travis, Dunlap "We've Got Our Eyes on You: When Surveillance by Unmanned Aircraft Systems Constitutes a Fourth Amendment Search" 2009 51(Fall) *South Texas Law Review* 173.

Warren, Samuel and Brandeis, Louis "The Right to Privacy" 1890 4 *Harvard Law Review* 193.

7   Drone terrorism
The ascent of evil

Be afraid. Be very afraid.

Geena Davis in The Fly

If 900g of weapons-grade anthrax was dropped from a drone at a height of 100m just upwind of a large city of 1.5 million people, all inhabitants would become infected. Even with the most aggressive medical measures that can realistically be taken during an epidemic, a study estimates that approximately 123,000 people would die—40 times more fatalities than from the 2001 World Trade Center terrorist attacks.

The chilling scenario above was one that was put forward more than a decade ago by Eugene Miasnikov in his report "Threat of Terrorism Using Unmanned Aerial Vehicles" (2005). If drones in the hands of terrorists back in 2005 caused a plausible threat, imagine the threat that exists today. As science and technological innovation continues to rampage we often lose sight of how much the world has changed—and in this instance, the extent to which terrorists will go in order to achieve their objectives. With this is mind, consider the following modern-day scenario.

A terrorist organization parks a small removal van in a crowded street of a major city under the flight path of a nearby international airport. The van's canopy has an open top but the sides are high and its payload of half a dozen high-performance quadcopter drones is obscured from the view of passers-by. To each drone is attached an explosive device—not dissimilar to those worn by suicide terrorists. The day and time chosen have been well planned to coincide with the runway being used for take-off. The targeted aircraft—an Airbus A380—is departing with a full payload of passengers and fuel, possibly in excess of 500 passengers and over 250 tonnes of fuel. The aircraft lifts off and the drones are launched remotely and rapidly ascend. With the aid of the high-resolution cameras onboard, the controllers are able to direct the drones into the path of the A380's four enormous engines. The catastrophic consequences are beyond belief and immediately impact upon the lives of the families and friends of the 500 fatalities and world economies are sent into chaos.

The situation described above is not inconceivable. Hoping that such a deplorable act upon humanity would never eventuate is no deterrent to the minds of terrorists seeking to inflict maximum carnage and media attention. As a society we should not allow ourselves to be detracted from the goal of safeguarding public security just because the thought of such acts is unconscionable. Previously inconceivable acts, such the 9–11 attacks, should now be prevented from occuring—and this will require a more sophisticated process of hazard identification. If such acts are conceivable to terrorists then they should be conceivable to those charged with the responsibility of public safety and security. In other words, when it comes to safety and security management, unknown risks should never equate to foreseeable risks—just because they were unforeseen.

Over the past half-century, terrorist attacks have progressively become the most evil hallmark of modern society. In complete defiance of human logic and moral canons the world has forever lost a significant element of its innocent creation. Threats from terrorist drones first became an issue of international concern following the September 11, 2001 attacks in New York City. Until then the use of commercial airliners hijacked by terrorists as weapons of mass destruction had eluded rational contemplation and traditional notions of reasonable foreseeability.

Creativity and innovation in the aviation industry has not always been directed toward achieving beneficial outcomes for society. In Chapter 4 we looked at the "good, bad and ugly" in terms of drone applications. In this chapter we now consider the "evil." A range of terrorist, insurgent, criminal, corporate and activist threat groups have already demonstrated the ability to use civilian drones for attacks and intelligence gathering. Such applications are far from being hypothetical propositions but, rather, pose a serious and vexing question: As drones become more commonplace in society, does the threat of them being used by terrorists as a weapon of destruction commensurately increase?

Drones have an obvious appeal to the extremist mind. Drones are difficult to detect, agile and capable of being controlled from afar and of flying into crowded or remote places. Drones can fly from practically anywhere into a sports stadium or a nuclear power plant. They can be affixed with explosives or chemical agents. And no one has to die to complete the mission—not that that is necessarily a concern of modern-day terrorists. Drones can combine the intimacy and stealth of a suicide bomber with the power and range of an armed aircraft.

An issue of international concern

With a twist of tragic irony far surpassing Shakespearian imagination, the threat of drone terrorism was born from the very desire to protect society. The rapid escalation in drone technology and capability was a direct result of strategies aimed at attacking terrorist targets. Retired Admiral Dennis Blair, who served as President Obama's first Director of National Intelligence, said he was concerned that the proliferation of armed drones—a potential outgrowth of the US reliance on drones to attack and kill terrorists—could well backfire. The US development and growing use of armed drones has "opened a huge Pandora's box which will make us wish we had never invented the drone."

Fortunately, there have thus far been very few instances of individual terrorists using drones to undertake attacks but those that have occurred have attracted worldwide media coverage. The international terrorist group Hezbollah, driven by resistance to Israel, has been flying drones for intelligence gathering into Israeli and Lebanese airspace for over a decade. In September 2014 Hezbollah carried out a "successful" drone strike, killing 23 Syrian "rebels."

Other terrorist groups using this new technology include Islamic State, who released videos showing drones being used for reconnaissance in Iraq. In 2013 a drone entered Israeli airspace and headed toward the coastal city of Ashdod. It was detected and then destroyed by a Patriot missile. The drone had been launched by the terrorist group Hamas who claim to have developed more than three types of military-style drones. Hamas promises more to come including some intended for "suicide missions." It has also been reported that Al-Qaeda has planned to use RPA for a range of brutal attacks.

Dennis Blair, as referenced above, portrays a disturbing prediction of what future terrorist acts may include:

> I do fear that if al Qaeda can develop a drone, its first thought will be to use it to kill our president, and senior officials and senior officers. It is possible, without a great deal of intelligence, to do something with a drone you cannot do with a high-powered rifle or driving a car full of explosives and other ways terrorists now use to try killing senior officials.

During the Obama administration, armed drone strikes accelerated against individuals and groups in Pakistan, Yemen, Somalia and Afghanistan, in a campaign that is almost entirely secret. The government is coming under increasing pressure to unveil at least some details of the secretive drone counter-terrorist campaign, which is carried out by the Joint Special Operations Command and by the CIA. It has been suggested that this strategy has only been partly thought through with the repercussions of its expanded drone attack campaign being the inevitable proliferation of drone technology to other countries including terrorist organizations.

Outside areas of civil unrest and war zones, there are increasing instances of home-grown drone terrorism. In 2012 the USA came under threat when a graduate student from Massachusetts plotted to strap plastic explosives to small drones and fly them into the Pentagon, the White House and the US Capitol building. In Japan it has been reported that a drone carrying a bottle of radioactive sand from Fukushima landed at the office of the Japanese Prime Minister in April 2015.

In the UK the Metropolitan Police has recorded over 30 suspicious drone-flying incidents around London between 2015 and 2016. Unidentified drones have also been flown over various landmarks in France, including the US Embassy and the Eiffel Tower. In 2016 at the Euro Cup qualifying match between Albania and Serbia the game was abandoned after a drone carrying a pro-Albanian banner was seen flying over the pitch. The incident caused brawls to break out between players, team officials and fans.

What is the scope of the drone terrorist threat?

Drones can be used as a threat to both national and international security. Malicious use of drones can inflict catastrophic consequences. They can carry chemical and biological weapons and disperse them in crowded areas. But other potential uses are unnerving: crop-dusting drones modified to disperse deadly chemicals, unmanned aircraft used as assassins and drones targeted to attack critical infrastructure. Furthermore, most existing air defense systems are in-effective against drones since they were developed to tackle threats of a different kind. Already, dozens of countries from Iran to China are using surveillance drones, and it is only a matter of time before swarms of armed drones, under the control of terrorists, take to the air.

An alarming report, "The Hostile Use of Drones" (Abbott *et al.*, 2016) was released in the UK in 2016 and warns that terrorists wanting to cause chaos, such as attacking nuclear power stations, have the potential to convert drones that are currently commercially available into flying armed missiles. The report suggests that the technology of remote control warfare is impossible to control.

A UK government counter-terrorism adviser, Detective Chief Inspector Colin Smith, has warned that terrorists could use commercially available drones to attack passenger planes. The security expert warned that small quadcopter drones could easily be used by terrorists for attacks and propaganda purposes. Terrorists could fly drones into an engine or load them with explosives to try to bring down a commercial airliner. Smith poses the question: "Are drone mitigation strategies going to be like the concrete bollards in front of airport terminals—something we can expect once the horse has bolted?"

The attractiveness of drones to terrorist groups has been well known for some time. Experts suggest (Gormley, 2003) that the attributes that make drones so attractive to terrorists include their ability to:

- attack targets that are difficult to reach by land (cars loaded with explosives or suicide terrorists);
- carry out a wide-scale (area) attack, aimed at inflicting a maximum death rate on a population (particularly, through the use of chemical or biological weapons in cities);
- provide covertness of attack preparation and flexibility in choice of a drone launch site;

- achieve a long range and acceptable accuracy with relatively inexpensive and increasingly available technology;
- defeat existing air defenses against targets such as low-flying drones;
- relative cost effectiveness of drones compared with ballistic missiles and manned airplanes;
- inflict a strong psychological effect on the community by scaring people and putting pressure on politicians (Miasnikov, 2005).

Recently in the US, the Department of Homeland Security issued a terror alert warning that drones could be used by terrorists to attack commercial aircraft after three drones were spotted in a single weekend in late 2015 flying above JFK International Airport. The sighting of the first drone was reported by the crew of a JetBlue flight arriving from Haiti. Just 2.5 hours later a Delta pilot, arriving at JFK from Orlando, reported a drone at approximately 1,400 ft and only 100 ft below the aircraft. The third report was from a Shuttle America flight arriving from Richmond, Virginia. And all this in the space of just two days.

The Homeland Security alert also warned:

> the rising trend in UAS incidents within the National Airspace System will continue, as UAS gain wider appeal with recreational users and commercial applications. While many of these encounters are not malicious in nature, they underscore potential security vulnerabilities . . . that could be used by adversaries to leverage UAS as part of an attack.

Anti-social application of technology

Every new technology has inspired anxieties about its effects; from comic books being categorized as "seduction of the innocent" to television as "the plugging drug." When Google introduced Google Earth the panic it first created included predictions of the widespread crime of voyeurism. Fears and anxieties in society often stem from fear of the unknown. A distinctive and undeniable feature of humanity is the desire to understand the world around us. This innate desire shakes even the most rational, intelligent thinkers—the thought that another human being could be interfering with the solitude of an individual without their knowledge.

Regardless of scale, human beings each have a right to security from others, reiterating the fine line between the protection of an individual's fundamental human rights and the control and harnessing of an unpredictable new technology.

We demand knowledge, we fear the unknown and when we receive an insufficient answer to the simplest questions (or more simply the answer we don't want to hear), we radiate energy of rejection. The emerging risks associated with drones are at the forefront of citizens' minds as they seek out any possible flaws before they choose to understand this unmistakably great innovation. Consequently, we are programmed to prioritize the option of rejection or prohibition before we realize the opportunity of betterment. The thought of the potential consequences of drones in the hands of terrorists further strengthens this view.

If acts of terrorism attracted any degree of moral fortitude it would more likely arise in the execution of direct attacks. History has proven this not to be the case. Given the physical remoteness from the real world in the drone-operating environment, any notions of conscious wrongdoing would be even less likely. This psychological disposition could give rise to a new wave of terrorists.

A person's detachment from physical reality might lead to decisions and actions that are inconsistent with the individual's normal morality, or that indulge fantasies that are normally kept under control by social norms. The detachment, combined with the thrill of live entertainment, can be expected to lead to enthusiastic voyeurism, which constitutes harassment, and will on occasions cross the boundary into stalking, and in some cases may culminate in acts of violence and even terrorism. The likelihood is that the social and institutional controls will be less effective than is the case in military contexts, and in the case of micro and nano-drones, perhaps almost entirely ineffective.

In conclusion, it can be said that the unprecedented growth in unmanned aircraft and the unbounded nature of all aircraft operations poses complex security challenges on a personal, corporate and national scale. The societal expectation of the maintenance of adequate security and privacy standards in a complex yet volatile world imposes even greater demands on governments and their regulatory agencies.

Combating the threat

Aviation is generally regarded as the most strictly and extensively regulated industry. It is therefore logical to conclude that the solution for controlling this new form of aircraft will be found in passing relevant laws and regulations. However, attempting to legislate against random acts of stupidity is difficult, particularly in the fast-moving world of technology. Also, "don't be an idiot" lacks legal clarity. Jonathan Rupprecht, a Florida-based lawyer specializing in unmanned aircraft, divides stupid drone owners into two groups, the "how high can it fly" group and the "I will fly it wherever I want" group. Obviously the latter grouping may also include acts of terrorism.

It is the freedom and agility with which aeronautical activities can readily transcend previously restrictive geographic and political boundaries that truly differentiates flying from all other modes of transport. To harness this freedom for the betterment of all, aviation regulation provides the requisite authority, responsibility and sanctions. The regulation of aerial activities is as fundamental and rudimentary to the aviation industry as civil order is to modern society. In no other field of human endeavor or branch of law does there exist such a vital yet symbiotic relationship.

As we have seen in previous chapters, international harmonization of aviation standards has been achieved through treaties. The Chicago Convention of 1944 is by far the most prolifically ratified international treaty. More than 190 sovereign states have ratified this convention and in so doing have agreed, under international air law, to be bound by the technical and operational standards developed by ICAO and as detailed in the 19 Annexes.

With respect to the control of drones, ICAO and other intergovernmental organizations such as the UN are struggling to keep pace with the rate of development of unmanned aircraft technology and their almost exponential rate of proliferation in society. The adoption of international aviation standards by contracting states have in most instances been extended by each state to also cover domestic aviation. However, as we saw in Chapter 4, ICAO is yet to promulgate international standards relating to unmanned aircraft. It has therefore been left up to sovereign states as to how (or if) they choose to regulate drone activities. And what makes this all the more concerning is that currently very few states have any restrictions on the purchase of drones. Buying a can of spray paint in many countries is infinitely more difficult than to purchase a potential weapon of mass destruction.

Conflict of laws and international disputes

In international aviation the concept of sovereignty is the keystone upon which virtually all air law is founded. It was World War I that brought about the realization of both the importance of aviation and its potential danger to states and their citizens in threatening their sovereignty. It was therefore not surprising that the first Article of the Paris Convention 1919, and also subsequently Article 1 of the Chicago Convention 1944, recognizes that: "every State has complete and exclusive sovereignty over the airspace above its territory."

International aviation systems and regulatory mechanisms are intrinsically linked and the need for equivalent levels of safety across the broad range of unmanned aircraft, from nano to microscopic drones to large reapers, is pertinent to the safety of all states and individuals. The rate of drone development and the commensurate potential for invasion of privacy, and the threat to national, corporate and personal security, has understandably generated considerable public debate.

International aviation law has delivered an impressive array of sources, including bilateral and multilateral treaties, intergovernmental organizations, non-government organizations and influential declarations aimed at the cooperative management of global problems. While all in some way contribute to the safety and efficiency of international air transportation, the maintaining of peace and security of individuals, states and the global sphere is the underlying fundamental. Yet with each power comes an additional list of restrictions and with each binding treaty a set of loopholes or ambiguities to ensure the protection of each state's autonomy and that no ultimate, definitive power can govern or possibly corrupt the entire global system. Nations are bound by international law only if they agree to be bound.

When an act of drone terrorism enters the international arena, the ability of governments to impose deterrent penalties—irrespective of their harshness—becomes all the more difficult. Consider the battle to bring to justice the perpetrators of the apparent missile attack on Malaysian Airlines international passenger flight MH17 from Amsterdam to Kuala Lumpur in July 2014. If—and it's a big if—an investigation finds Russia and/or the separatists responsible it will be a massive challenge to bring the matter to a court, let alone secure a conviction.

As with any international incident involving multiple nations and their varying legal systems, the International Court of Justice (ICJ) and the International Criminal Court (ICC) would be the appropriate forums to hear such a matter. But in the case of MH17 neither Russia nor the Ukraine accept the compulsory jurisdiction of the ICJ. Neither country has ratified the treaty that established the ICC, meaning that its citizens are not compelled to appear before the court.

The UN Security Council can force any matter to be investigated by the ICC, regardless of whether a nation state has ratified its statutes or not, yet Russia is one of five permanent member countries that has veto power in the Council. Furthermore, while a degree of restitution may be attained against offending nations through economic sanctions, diplomatic pressure and sustained adverse media coverage—such measures would be prove to be futile against terrorist organizations.

As has been discussed above, in the case of unmanned aircraft, international treaties are yet to provide the mechanism for individual governments, and their law enforcement agencies, to effectively control drones domestically. In the following section we will look at other measures that may be considered to mitigate the risk of drone terrorist activities.

Compulsory registration of drones

As drones become more common, many governments are considering a number of options to restrict their use. Registration of drones, as with cars, airplanes or even guns, is now being introduced all over the world with the FAA leading the way, and over 500,000 drones were registered in the first few months of October 2015. It has also been suggested that drone controllers should be subjected, at a minimum, to the same background check standards as persons granted unescorted access to security restricted areas of airports as is required under ICAO Annex 17.

The UK and Australia are also building similar registration systems to follow suit. It's far from clear how registration would mitigate an act of terrorism, as it is more of a system for tracking law-abiding citizens' drones. David Dunn (2016), Professor of International Politics at Birmingham University, believes that any licensing system is unlikely to deter terrorists:

Law abiding citizens are likely to register, but it would be very difficult to stop terrorists and other criminals from purchasing drones abroad and then using them here. Up until now it was expensive and required skill to be able to fly an aircraft—which acted as a form of regulation in itself. Now, you can fly these things relatively easily over people's heads.

In the UK the House of Lords has called upon the EU to introduce a compulsory registration system for the devices, but the plans have stalled. Drone owners currently don't have to register their devices in the UK, but operators need permission from the British CAA to fly them for commercial purposes or over long distances. Currently in the UK, anyone can own and operate a drone for non-commercial purposes that weighs less than 20kg (3st 2lb).

Mitigating the drone terrorist threat?

As we have seen above, it is obvious that legislative restrictions alone on the use of drones would in most instances prove to be futile when it comes to acts of drone-related terrorism. There has been very little indication that governments are prepared to prohibit the importation or manufacture of drones or even of limiting the payload capacity of commercial drones that are sold. Further complicating this issue is the fact that, in many instances, drones are purchased online.

Creating a greater awareness in the broader community of the extent to which drones may be used by terrorists (and other criminals) including publicizing the dangers—without hysterics—may be a good start. Also, manufacturers and distributors of drones and training establishments throughout the world should be more vigilant of the possible use of drones for terrorist activities. By way of parallel, many governments have passed legislation requiring retailers of chlorine (for swimming pools) and household fertilizers to report certain sales or suspicious transactions.

International arrangements regulating the export of drone technology could be refined and strengthened with terrorist activities in mind, with special attention on drones equipped with technologies that can evade radar or have high-performance capabilities.

While the rapid advancement of drone technological development has created the problem it may also provide the solution. By far the most effective method of protecting targets from drone attacks may be with the installation (or possibly mandating) of geo-fencing or g-gate technology software. Pre-programing geo-fencing areas would mean that drones would be automatically shut down if they tried to enter certain sites. NASA is also currently working on a tracking system but a working prototype is not expected until 2019.

Drone manufacturers could be required to install the GPS coordinates of government-mandated no-fly zones and have drones automatically shut down if they approach such a space. DJI, the world's largest commercial drone-maker, is one of the leaders in geo-fencing technology. With drone sales in excess of

US$1 billion in 2015, it recently released its geo-fencing software to restrict drones from flying near aerodromes and other restricted areas on a worldwide basis.

The drones will no longer be able to fly near wildfires, prisons, power plants, near professional sporting events or areas the US president is visiting. It is proposed that all DJI drones will have the software installed by default. In practice, this means that drones will not be able to enter into, take-off or land in restricted areas. The software will automatically update with new information on restrictions, meaning drones will be able to respond to changing environments such as areas of natural disasters or one-off sporting events.

Other technological defenses against the hostile use of drones are with the installation of security alert systems when drones appear in no-fly zones. One American company—DroneShield—has been awarded contracts to protect certain locations from possible terrorist attacks including the Boston Marathon. It is likely that this technology will be increasingly utilized in security-sensitive sites and restricted areas.

In the UK the Remote Control Project, run by the Oxford Research Group, has called on the British government to fund the development of military-style lasers to shoot drones down and the creation of jamming and early-warning systems to be used by police. But such devices would require amendment of UK laws over the use of such jammers.

Laser technology to destroy drones in many instances has failed to live up to expectations, either struggling to stay fully powered for long periods or being disrupted by dust and fog. However, in the US, Boeing has unveiled its new laser-powered anti-drone technology. The Compact Laser Weapons System is a portable, tripod-mounted device armed with a high-powered laser that can destroy a quadcopter drone in a matter of seconds. The system is relatively inexpensive to operate and features an unlimited magazine, which means many drones can be destroyed. However, this system will not be available for a few more years.

A final word

With the continually increasing rate of terrorist attacks around the world, most nations have passed laws that prohibit and punish those conspiring and participating in terrorist activities. Obviously increased sanctions and penalties relating to acts of terrorism generally may assist to deter drone-related activities, as would specific referencing to the use of drones. Terrorist mitigation strategies beyond the efforts of governments may be required as Miasnikov (2005) suggested in his report:

> Terrorist activity can be prevented only through the coordinated efforts of the government and civil society. The government cannot efficiently fight terrorists without the active involvement of the population. The first step toward creating such an alliance is to recognize the threat and its potential consequences.

It is hoped that this chapter has assisted in raising awareness of the threat of drone-related terrorist activities. However, recognizing the threat and potential consequences of terrorists' drones today should not be such a difficult task. Perhaps a decade ago—and when civilian usage of drones was rather uncommon and the aircraft less sophisticated—some degree of clairvoyance was required. Not so today. All that is required is to switch on the television and watch the news.

References

Abbott, Chris, Clarke, Matthew, Hathorn, Steve and Hickie, Scott "The Hostile Use of Drones by Non-state Actors against British Targets" Oxford Research Group, published by Remote Control Project, London, UK, January 2016, p. 11.

Bartsch, Ron *International Aviation Law* Ashgate, Farnham, UK, 2012.

Bartsch, Ron "To Catch a Drone: Security and Privacy Challenges in a High-Tech Age" Parliament House Lecture, September 24, 2014, Canberra, Australia, www.aph.gov.au/About_Parliament/Parliamentary_Departments/Parliamentary_Library/pubs/Vis/vis1415, accessed August 15, 2016.

Cho, George "Unmanned Aerial Vehicles: Emerging Policy and Regulatory Issues" (2013) 22(2) *Journal of Law, Information and Science* 201.

Dunn, David "Warning over Drones Use by Terrorists" BBC News, January 11, 2016, www.bbc.com/news/technology-35280402, accessed August 15, 2016.

Gormley, Dennis "UAVs and Cruise Missiles as Possible Terrorist Weapons" in *New Challenges in Missile Proliferation, Missile Defences, and Space Security*, Ed. James Clay Moltz, Occasional Paper No. 12, Center for Nonproliferation Studies and Mauntbatten Centre for International Studies, July 2003.

International Civil Aviation Authority (ICAO) Annex 17—Security—Safeguarding International Civil Aviation against Acts of Unlawful Interference. Paragraph 4.2.4 refers to standards for security checks for airside personnel.

Miasnikov, Eugene "Threat of Terrorism Using Unmanned Aerial Vehicles" Center for Arms Control, Energy and Environmental Studies Moscow Institute of Physics and Technology, June 2004, 26 pages. Translated into English March 2005.

8   To err is human

Human factors

I always find beauty in things that are odd and imperfect—they are much more interesting.

Marc Jacobs

For all of the opportunity that drone technology offers, humans still need to be involved. They are in control of the flight, manufacture and maintenance of this wonderful new gadget. And humans are imperfect creatures. We will make mistakes, have limits and just plain stuff up. We will continue to do so when it comes to our unmanned aircraft. It is only through accepting our human weakness and learning from it that we can improve and work around our fallibility. This is the focus of human factors.

Many of the human factor principles for aviation were identified through the investigation of accidents and incidents—an approach sometimes referred to as "tombstone safety" (Hobbs and Shively, 2013). Thankfully, for unmanned aircraft, much of the core learning has been done . . . the industry now has a strong belief in managing human factors, and a solid body of knowledge to know where to look in the UAS to start. Sweetly for researchers, the cockpit is also now accessible for active observation. It all bodes well for a safer, less "terminal" approach to improving safety in this generation of aircraft.

So then, how will pilots stuff up? What are their limitations? What is being done about it?

(Hawkins 1995) lists the classic human errors as design induced and operator induced; random, systematic and sporadic; by omission, commission or substitution; and reversible (the good ones) and irreversible (oh no!).

Human factor issues are then considered in terms of Hawkins (1995):

*   fatigue, body rhythms and sleep;
*   fitness and performance (of humans);
*   vision and visual illusions;
*   motivation and leadership (no difference for unmanned aircraft);
*   communication: language and speech;
*   attitudes and persuasion;

- training and training devices;
- documentation;
- displays and controls; and
- space and layout.

iPhones versus desktops

Let's start with the displays and controls; the part the pilot interacts with to control the aircraft.

The ground stations vary widely in their technology, flexibility, layout and design, as demonstrated by the three examples outlined below. From a full room set up, to a mobile phone and of course everything that comes in between.

The Parrot aircraft series have a dedicated controller with screen and integrated controls (Skycontroller, 2016).

Meanwhile the DGI Phantom works with existing iPhones and iPads and has "Fly with touch." Tap the screen of your iPhone or iPad to move in a particular direction; keep flying by tapping elsewhere on the screen. It also has "Visual tracking" (fly while keeping the camera pointed steadily at a moving subject with fully automated tracking) (Apple Shop, 2016).

Your iPhone presents the view from the aircraft (Apple Shop, 2016) and you can tap a building on the display to fly toward it, then tap the next building, then the next building and so you fly truly visually (Apple Shop, 2016). Or select your chosen object and actively track the object; it recognizes your subject, follows them automatically and keeps them in the frame. No GPS bracelet, tracker or beacon required (DJI, 2016). This adaptation of the controls means you can fly around as easily as you would walk around your own backyard.

Getting up to the heavy end of aircraft, the Global Hawk (with a wingspan that equals a Boeing 737) still needs a complex control station, with multiple screen displays that are reminiscent of a full-sized cockpit, but include the visuals by video feed instead of a window (NASA, 2016).

To date, the control stations of many unmanned aircraft have had less-than-ideal human-system interfaces. This isn't surprising—the technology is springing up from toy manufacturers, homebuilders and robotics labs, as well as traditional aircraft manufacturers. Many of those building the aircraft have no idea of existing human factor principles, and could improve design if they applied existing knowledge.

> On the other hand, the design problems may stem from the simple lack of suitable guidance material.
>
> (Hobbs and Shivley, 2013)

NASA has released draft guidance material for the ground control station design (Hobbs and Lyall, 2015). This guidance material examines the special considerations of UAS:

A. Loss of natural sensing
B. Control and communication via radio link
C. Physical characteristics of the control station
D. In-flight transfer of control
E. Unique flight characteristics of unmanned aircraft
F. Flight termination
G. Reliance on automation
H. Widespread use of interfaces based on consumer products

(Hobbs and Lyall, 2015)

The great outdoors

Space and layout for the cockpit are quite markedly different for the pilot of an unmanned aircraft. The nicest change you might notice is that the pilots and their controls don't have to be squashed into an aircraft cockpit any more. Space and room are a luxury in an aircraft cockpit.

The new cockpit has plenty of space; you can even bring your own chair. Sit outside, in your car, on a picnic blanket, by/in the pool even. But it isn't necessarily air conditioned (or heated), shaded, quiet or dry. This leads to increased pressure on the pilot, if not managed. Nobody who is cold, wet and windblown is going to be giving full attention to the job.

Paperwork, paperwork and yet more paperwork

Another interesting issue is the matter of the aircraft's documentation. Conventional aircraft have suites of checklists and manuals to deal with anything that can go wrong on the aircraft—that's what is in those big black briefcases pilots lug onto each flight. Many pilots now carry much of the information electronically, in an "electronic flight bag," which is considerably lighter.

UA pilots also don't have to carry their documents in physical form, and in fact many operators of smaller drones carry them electronically, accessed via their tablet or phone. While the likes of the pilots of a Global Hawk will have a tried and tested approach to their emergency situations, and the smaller drone operators' standard emergency approach is to simply land the aircraft, somewhere in between there is a switch over. It remains to be seen if accessing the required detailed procedures and checklists on the same tablet the pilot is using to steer the aircraft will cause problems if the pilot can only do one of the two tasks effectively.

Time to get new glasses

Our eyes are a stand out issue. Our eyes are well renowned for being varied in their capability, while having inbuilt limitations across the board. The limits of our vision and happenstance of visual illusions can frustrate us all at some time.

An aircraft pilot has to undertake vision testing and limits are placed on the degree to which this can be impaired (so they can distinguish the likes of emergency lights and landing aids, and see sufficient width clearly).

An unmanned aircraft pilot can dive on in to flying not only without any visual testing, but also potentially unaware of their own personal sight issues such as color blindness, tunnel vision or limited range of sight.

Should the ground station be immune to variability in our eyesight? Or should the pilots have minimum eyesight requirements to be allowed to fly?

If the aircraft controller displays in grayscale, is color blindness still a limiting factor? If the spatial view is through the controller alone (BVLOS), the needed vision is to a narrowed view. Is tunnel vision such an issue any more?

And eyes are not the only rich sensory cues available to the pilot of a conventional aircraft—who also has auditory, proprioceptive and olfactory sensations. The absence of these cues in a UAS makes it more difficult for the pilot to maintain an awareness of the aircraft's state (Williams, 2006).

On the counterpoint, making up for all of these cues in one simple ground control can lead to significant information and visual overload and itself create a problem—and that is a difficult balancing act with the range of people flying drones. Traditional jet cockpit displays would be beyond usable for a new aviator who is also required to maintain sight of the drone. However, for an experienced aviator, the display is presenting many of the cues normally available in flight and would improve control and awareness for the pilot.

Depth perception is also a significant sense input change. Depth perception is the ability to see the distance of an object, particularly in three dimensions, relative to other objects.

The term depth perception refers to our ability to determine distances between objects and to see the world in three dimensions. To do this accurately, you must have binocular stereoscopic vision (stereopsis).

For example, flying a kite over the top of a playground. The further you are from the kite, and the higher the kite flies, the harder it is to assess whether the kite is directly over the top of the playground. Therefore it becomes more difficult to know if the kite is a danger to the children in the playground if the wind should drop. Test your own depth perception at www.mediacollege.com/3d/depth-perception/test.html (as at May 15, 2016).

That depth perception problem is exacerbated if you happen to be controlling the kite by watching a screen, where stereoscopic vision of the kite isn't available, and on top of that you don't have the string connecting you—to zing, wriggle and tug in your hand and let you know something is up.

If someone lacks stereopsis, they are forced to rely on other visual cues to gauge depth, and their depth perception will be less accurate (EyeHealthWeb, 2016).

There is some debate raging about how severe that "less accurate" is—those who have lost one eye, at nearly any stage in life, rapidly relearn depth perception without the stereopsis (and perceive all movies in 3D, unlike the rest of us who have to pay to see the 3D version and wear the super cute glasses). It may be most pronounced when someone with two eyes loses that stereopsis skill and hasn't yet trained their brain to the other cues used by those without the binary vision. But does that mean that you can train your brain to perceive depth just as well, or if not as well, *nearly as well*? Does that also mean then that you can improve your depth perception with training?

Then there is also the question of screen usage in the outdoors, where most small drones are operated. Where is the sun, how bright is it today, are there reflections, glare, shadows and screen contrast? There is also the familiar problem of reflected glare (ControlVision, 2016).

Hit the gym

Fitness and performance have long been known to correlate. But, at a personal level, with long-term exposure to the effects of the high altitudes required for jet flight, even small aircraft flight is also known to effect pilots. The effect is mixed; the body's respiratory and circulatory system gears up to adapt but the drier, ozone-loaded air can cause the body to suffer dehydration and minor toxicity issues (Hawkins, 1995).

For pilots, there is a distinct personal health opportunity in spending less time at altitude, so flights controlled from the ground are a personal health benefit. Oh, and for pilots who struggle with (or are happy with) a nicotine addiction, their cigarette is far easier to access when you can hand over control and walk away from the whole aircraft for five minutes.

However, hours spent hunched over a controller may have inherent risks. Could full-time piloting of an unmanned aircraft have as significant an effect as heavy video gaming? Long-term screen time has been demonstrated to increase obesity, tendonitis, RSI and possibly increase the risk of epileptic fits (Subrahmanyam *et al.*, 2000; Griffiths, 2005; CAMI 2016).

Jet lag keeping you up?

The issues of jet lag are well understood by anyone who has undertaken a significant time zone shift. There is nothing that is so much fun as lying in bed wide awake and hungry at 2am having just shifted time zones by eight hours or more. Fatigue, body rhythms and sleep quality make a difference in all of us.

For pilots, working shifts in different time zones, circadian rhythm disruption, fatigue, sleep pattern disruption and biological rhythms (food, medicine, etc.) disruption are all part and parcel of the challenges of their job. This could conceivably be a non-event in the not too distant future, if they are able to fly the aircraft from their "home base" and stay in a natural rhythm or, even better, fly the long-haul flight for a standard working day and hand over to the pilot further around the globe for the next leg, which is during that pilot's standard working day.

Long-haul flights have already been undertaken. QinetiQs military Zephyr, which continues to break world records for airborne flight time, at the time of publication has completed a fully airborne14-day flight (QinetiQ, 2010). Take a moment to think about the pilots of that flight—there was more than one of them. If they were all in one location (safer handover if you can see each other), then some of them would have been flying the aircraft in the aircraft's "night" while they were in daytime, and vice versa. What sort of effect does that have on the pilots, their alertness and their body rhythms?

And there's also boredom . . .

Car drivers traveling long distances have long been advised to take rest breaks, in recognition of the fact that the activity of "staring," while actively engaged, is still mind numbing and there's only so much you can do in one day. Fatigue is even measured for us as having an impact on reaction times equivalent to levels of alcohol in the system.

The EU has guidelines recommending all drivers take a 45-minute break for every 4.5 hours driving (European Union Transport, 2016). The UK recommends drivers take regular 15-minute breaks in journeys over three hours, and should aim to stop every two hours or so (Automobile Association (UK), 2011). Australia recommends a break every two hours. Many cars and GPSs are even fitted with break warnings, to remind you to stop and stretch.

Commercial pilots have strict rest and flying time rules in place. However, private pilots have more lenience. Drone pilots are not subject to any rest rules, and can fly as required. For many drones, with battery lives sufficient to fly one to two hours, it really isn't an issue. But consider that multiday Zephyr flight. Imagine staring at the world 60,000 or more feet below for seven, eight or nine hours straight, via a video screen. The Zephyr team do rotate pilots.

A challenge for the designer of the ground control station is to maintain enough activity for the pilot to be engaged during extended periods of low workload, but also allow that the pilot must be prepared for the possibility that workload may increase rapidly (Hobbs and Shivley, 2013).

> The Massachusetts Institute of Technology (MIT) is also studying the issue of boredom on drone pilots.
>
> On its surface, operating a military drone looks a lot like playing a video game: Operators sit at workstations, manipulating joysticks to remotely adjust a drone's pitch and elevation, while grainy images from the vehicle's camera project onto a computer screen. An operator can issue a command to fire if an image reveals a hostile target, but such adrenaline-charged moments are few and far between.
>
> (Chu, 2012)

Mary Cummings (Chu, 2012), Associate Professor of Aeronautics and Astronautics and Engineering Systems at MIT, refers to the piloting of an unmanned aircraft as "a human is effectively babysitting the automation."

MIT's research has found that, at some level, a degree of intentional distraction can improve the general performance of the pilots. This is the sort of research that couldn't reasonably be done in a live commercial flight (hey yeah, distract my pilot as we fly in the region, upwind of an active volcano, great idea), but it has application back into the commercial flights and could lead to improvements in safety (Chu, 2012).

And apply the idea at home—if we liken it to driving with children in the car. There is an optimum level of distraction from them, all you have to do is find it and train them to do it for you (we wish you luck).

Are you listening to me?

Until we are mind readers, we will experience communication issues—misunderstanding, misinterpretation and just plain not hearing (or not listening).

More extended flight capabilities of UAS rely on a significant communication load, between observers and the pilot, and pilots when handing over control, as well as the normal air traffic control communications (radio monitoring and transmissions if needed). This is further hampered by delays in the system. Consider an observer noticing and reporting something, radio retransmissions from air traffic control to the ground observers and lag time in the controls, where an input to change flight direction isn't straight into the aircraft (hard right!), but must be transmitted to the aircraft first (I'd like you to turn hard right!).

Procedures and protocols, like those used with aircraft in heavy traffic areas around airports, can go a long way to managing the risks and issues around these communications. But each flight team needs to establish and be naturally familiar with the communication "norms" for this to work. In general aviation, students learn the normal language they will need during training. For the likes of unmanned aircraft operators handing over control to the other side of the world for a flight, and observers handing VLOS over to another observer as an unmanned aircraft crests a hill, the language isn't normalized yet. Moving from one team to the next will change the communications.

Somewhere, someday, someone is going to say turn right and the unmanned aircraft pilot is going to turn left. It's probably already happened and nothing at all went wrong. Or maybe the drone hit the tree and it cost them a few thousand dollars in equipment.

Back off, I'm happy with my bad mood

The pilot's attitude, to safety, professionalism, risk and values, all affect his or her approach to the flight and any issues that come up.

Although no lives are at stake onboard the unmanned aircraft, the pilot is still responsible for the protection of life and property on the ground (Hobbs and Lyall, 2015) and other airspace users. But there is a significant difference between crashing a vehicle when you are onboard, and crashing a vehicle that's 500m away and can't hit you.

Take the example of the gentleman who crashed his drone into the Sydney Harbour Bridge. He reportedly lost control of his drone, then lost sight of it and assumed it had crashed into the water. So he simply packed up his kit and moved on. There was no personal risk perceived. However, the out-of-control drone ended up in an operational train area on the Sydney Harbour Bridge and certainly presented a risk.

Did it crash because he was less worried than a pilot onboard an aircraft would be? Or did it crash anyway, and he was less upset because he didn't see it happen?

On the other hand, for the dangerous, dull and dirty jobs, if you don't have to sit in the aircraft all afternoon, putting your life at risk, would you be more willing to do the work?

I didn't know there was a rule for that

Another issue new to the human factors work for UAs is the matter of willingness to comply with the rules and normalized behaviors of aviation. A conventional aircraft pilot must be licensed and flies an aircraft that is clearly registered and has radio contact. The pilot's personal accountability is pretty tightly wrapped up. That said, aviation already experiences instances of "low flying" where pilots make illegal low approaches, for whatever reason (the good camera shot, or to see their own house up close perhaps?). These instances happen below radar coverage away from airports, and if there isn't another aircraft around at the time, they can be difficult to identify and prosecute.

How much freer then are unmanned aircraft pilots? They are able to commence flying without a license, registrations or physical connection to the aircraft—although not necessarily legally. The aircraft truly can crash into a house and the pilot can walk away unscathed but also *unknown*.

This level of anonymity may well lead to a reduced level of voluntary compliance with operating rules. Such a challenge with anonymity and breaking the rules isn't unusual to regulatory agencies. It is *relatively new* to aviation regulatory agencies. But there are precedents around, so aviation regulators can take lessons and experience from the techniques employed by the local police force who regularly manage our varied willingness to comply with speed limits, parking rules and car maintenance levels.

Note that the unmanned aircraft registration factor is being considered to some degree, with a number of countries implementing a range of registration rules (although the registration still isn't readable on the side of that 10cm tall aircraft from 30m away).

I learned good

Last but not least, there is the issue of training. Pilots in conventional aircraft receive training through formal processes, on a competency basis, which is regulated.

Pilots of UAs can technically start flying in most countries by treating their aircraft as a model aircraft, with no training in operation, emergency procedures, proximity rules or even safety. The training that is available can also be quite varied, and the training, where it is required, varies worldwide. Training is usually conducted with the new aircraft itself, so skills are aircraft, controller and software specific. It is not as easy as you might think, and numerous drones make only one flight at the hands of learners. Two nice examples on the web are http://dronenews.net/videos/someone-found-out-flying-a-drone-is-not-that-easy/ and www.youtube.com/watch?v=1Bv29pPcRhs (as at May 15, 2016).

Rest assured there are also plenty of training videos on the internet too. Training is one area of operations of UAS that will gain significantly from aviation's tradition of sharing bad experiences and lessons learnt to support research and education. At least one of the authors of this book has a drone still boxed up in the cupboard—afraid that learning to use the aircraft will cause the loss of the aircraft.

Want to know more?

We suggest following NASA's research, generally by Professor Alan Hobbs.

On a tangential note

Our research unearthed this illuminating article from War History Online (2015), sharing some "learnt" random and unusual considerations for human factors (published unedited, with permission).

> In August 1956, a drone went rogue over Southern California, threatening cities, including Los Angeles. In its aftermath, over 1,000 acres were destroyed, forests and scrub land were set ablaze while homes and property were damaged. But it wasn't because of the drone.
>
> With Cold War tensions on the rise and everyone fearing an invasion of the Red Menace, the US Navy began research on surface-to-air and air-to-air missiles. To test those missiles, they'd launch remote-controlled planes (drones) into the sky for target practice, mostly from the Naval Air Station Point Mugu in Ventura County.
>
> On 16 August 1956, a Grumman F6F-5K drone (also known as the WWII Hellcat) sat on one of their runways, painted bright red for extra visibility. At 11:34 AM, radio controllers launched it into the air, then sent it south toward the Pacific Ocean so the navy could shoot it down.

But the drone had other ideas. It veered to the left and continued to climb, heading southeast toward Los Angeles. The operators had lost control.

Enter Oxnard Air Force Base. The 437th Fighter–Interceptor responded by sending out two of its Northrop F–89D Scorpion twin–jet interceptors. The first jet-powered planes that could handle all types of weather, they were the most advanced aircraft of their time. They were also the first with computer-guided missiles, so what could possibly go wrong?

Manning the first Scorpion were First Lieutenant Hans Einstein, the pilot, and First Lt. CD Murray, the radar observer. Following them were First Lt. Richard Hurliman and First Lt. Walter Hale. Their orders were simple: take down the drone before it crashed into a populated area.

They reached the Hellcat to the northeast of Los Angeles at 30,000 feet. In response, it veered southwest, flying directly over the city before banking northwest toward the Santa Clara River Valley. It continued on till Santa Paula, the "Citrus Capital of the World," and slowly circled above the city.

The pilots waited, praying it wouldn't crash into the buildings below, hoping it would veer off again toward the countryside. Their prayers were answered. The drone headed northeast, zoomed over the city of Fillmore, then over the suburb of Frazier Park, and made a beeline toward Antelope Valley.

The pilots heaved a sigh of relief. They could finally take the thing down with their Mk 4 Folding-Fin Aerial Rockets (FFAR), also called the Mighty Mouse Rockets. The Scorpions were not equipped with machine guns or cannons. They had two options by which to fire them with: use their computers or fire them manually. Because the drone was so erratic, they chose the former.

The Scorpions had the new Hughes E-6 fire control system and AN/APG-40 radar, which were guided by an attack-plotting computer. All the pilots had to do was get the drone in their sights and fire. However the Hellcat chose to zig or zag, the computer would take care of it. So they fired.

And nothing happened. So they tried again. Nada. A design flaw in the fire control system wouldn't launch the rockets.

No problem. Switch to Plan B and fire their rockets manually. So they looked into their gunsights to aim . . . and discovered that they had a problem. No gunsights. To put in the new state-of-the-art Hughes E-6 system, the installers had to remove the gunsights.

So now they had to aim without any sights at a flying object that kept changing course while they themselves were in constant motion too. Fortunately, each F-89D came armed with 104 rockets each, and they only needed one to bring the drone down. What could possibly go wrong?

Bored with Antelope Valley, the drone veered south, heading back toward Los Angeles. As it passed over rural Castaic to the north of the city, the pilots fired 42 rockets. They passed beneath the drone, a few whacked the underside of the plane's fuselage, but none detonated.

Offended, the drone headed toward the town of Newhall. The interceptors replied by firing 64 rockets . . . none of which made a hit. The Rogue Drone veered away again, this time northwest toward the city of Palmdale. The interceptors fired their remaining 30 rockets . . . still no hit.

The interceptors had run out of missiles – all 208 of them. They were also running out of fuel, forcing them to go back to Oxnard.

The Hellcat had finally made up its mind. Running out of fuel, it flew toward Palmdale. The Palmdale Regional Airport came in sight as it ran out completely – thus sending it on a downward spiral of doom.

It cut through three Southern California Edison electric cables along Avenue P some eight miles from the airport. Its right wing hit the sand first. It flipped over, smashed into the desert, did several cartwheels before slamming into the ground where it flew apart into so many pieces, it was only in 1997 that archeologists found the remains.

Although the pilots hit nothing in the air, the ground was another thing completely. Fifteen rockets didn't go off but the rest did. Castaic saw 150 acres go up in flames. At Placerita Canyon, one rocket bounced along the ground, setting off a series of fires, one of which ignited some of the Indian Oil Co.'s oil sumps. At Soledad Canyon, 350 acres were set ablaze. Newhall wasn't hit, but it had to deal with the smoke.

The Hellcat hadn't hit Palmdale, but some of the rockets had, starting fires in and around the city. A piece of shrapnel shot through the window of Edna Carlson's house, bounced off her ceiling, burst through a wall, and finally parked inside one of her kitchen cabinets. More fragments punched through JR Hingle's garage, shot through his living room, and nearly hit his wife.

Larry Kempton was driving on Palmdale Blvd. with his mother when a rocket exploded in front of his car. It took out his left tire, radiator, hood, and windshield. At Placerita Canyon, two men had just left their truck to have lunch when a rocket blew it up.

It took two days and 500 firefighters to put all the fires out, and explosive ordnance disposal teams to remove 13 duds around Palmdale. Incredibly, no one was seriously hurt or killed.

The Battle of Palmdale confronted the US Air Force with an embarrassing fact – that two of their latest, state-of-art interceptors couldn't bring down an obsolete, pilot-less propeller-driven plane.

California faced a red menace that day, but it wasn't communist.

References

Apple Shop www.apple.com/shop/product/HJWF2LL/A/dji-phantom-4-camera-drone?fnode=a3&fs=f%3Ddrone%26fh%3D3603%252B45e6, April 15, 2016.

Automobile Association (UK) www.theaa.com/public_affairs/reports/tired-drivers. html, December 8, 2011, accessed April 16, 2016.

Chu, Jennifer "Boredom and Unmanned Aerial Vehicles" MIT News, November 14, 2012.

Civil Aerospace Medical Institute (CAMI) "Introduction to Aviation Physiology", FAA, 2016, www.faa.gov/pilots/training/airman_education/media/IntroAviationPhys.pdf, accessed April 16, 2016.

Control Vision http://store.controlvision.com/images/products/septa/ipad-reflection LR.jpg, 2016, accessed April 16, 2016.

DJI www.dji.com/product/phantom-4, 2016, accessed April 15, 2016.

European Union Transport http://ec.europa.eu/transport/modes/road/social_ provisions/driving_time/index_en.htm, March 3, 2016, accessed April 16, 2016.

EyeHealthWeb www.eyehealthweb.com/depth-perception/, 2016, accessed April 15, 2016.

Griffiths Mark "Video Games and Health: Video Gaming is Safe for Most Players and Can be Useful in Health Care" 2005, 331(7509) *British Medical Journal*, 122–123.

Hawkins, Frank H *Human Factors in Flight* Avebury Aviation, Aldershot, UK, 1995.

Hobbs, Alan and Lyall, Beth "Human Factors Guidelines for Unmanned Aircraft System Ground Control Stations" NASA, September 2015.

Hobbs, Alan and Shivley, Jay R "Human Factors Guidelines For UAS In The National Airspace System" *Proceedings of Association for Unmanned Vehicle Systems International* (AUVSI), Washington DC, August 12–15, 2013.

NASA www.nasa.gov/sites/default/files/674934main_controls.jpg, 2016, accessed April 15, 2016.

QinetiQ Press Release "After 14 Nights in the Air, QinetiQ Prepares to Land its Zephyr Solar Powered Unmanned Aircraft," July 23, 2010.

Skycontroller www.parrot.com/au/products/skycontroller/, 2016, accessed April 15, 2016.

Subrahmanyam, Kaveri, Kraut, Robert E, Greenfield, Patricia M and Gross, Elisheva F "The Impact of Home Computer Use on Children's Activities and Development" (2000) 10(2) *Children and Computer Technology*.

War History Online, "Two Jet Fighters Sent Up to Shoot Down a WWII Warbird in 1956 – Blasted 208 Rockets at it, it Survived" www.warhistoryonline.com/ military-vehicle-news, December 14, 2015, accessed May 10, 2016.

Williams, KW "Human Factors Implications of Unmanned Aircraft Accidents: Flight Control Problems," in *Human Factors of Remotely Operated Vehicles*, Ed. NJ Cooke, HL Pringle, HK Pedersen, and O. Connor (pp. 105–116), Elsevier, San Diego, CA. 2006

9   What's hot and happening with Remotely Piloted Aircraft Systems

> Technology gives us power, but it does not and cannot tell us how to use that power. Thanks to technology, we can instantly communicate across the world, but it still doesn't help us know what to say.
>
> Jonathan Sacks (a British rabbi, philosopher and scholar of Judaism)

When did the "flyover shot" of your wedding stop being the bane of Hollywood stars, and become a "must have" for any modern bride? What might be the next normal? What else is going on out there in the UAV industry that might just spark a creative industry off next year?

Flying alongside manned aircraft

If anyone dares to suggest that unmanned aircraft will never fly in the same space as manned aircraft, you can confidently tell them they are wrong. Paul Herrmann (Herrmann, 2016), Chief Pilot at Aerosonde Pty Ltd, explains that Aerosonde are already stretching the boundaries of operational norms by operating internationally, in controlled airspace (that is, with manned aircraft).

Aerosonde (Aerosonde, 2016) is a strategic business of AAI Corporation, an operating unit of Textron Systems, a Textron Inc. company. They provide small UAS capabilities to the military and scientific communities.

Their "integrated" operations include:

- Antarctica, Terra Nova Bay: supporting work in the region. They had to deal with more than 70kt winds, air temperatures below 40°C and icing. On the easier side, there was little air traffic, and the air traffic they are integrated with are all sophisticated and subject to airspace control (Herrmann, 2016).
- Taiwan: ocean flights out from Taiwan, in controlled airspace. The operation had to deal with moderate to severe turbulence and rain. These were high-altitude and long-range flights. For this operation, transponders were utilized, and VHF radio communications were direct to air traffic

control. Regional general aviation traffic was very limited in the flight area (Herrmann, 2016).

- Australia, Thevenard Island: flights integrated with all local traffic—charter, regular public transport and helicopters. The aircraft flew with a transponder, and operated as an IFR flight in Class G airspace (Herrmann, 2016).

Their lessons? There is a lot of effort involved to manage the "underlying see and avoid deficit." Danger and restricted areas can be used to manage the risks, but were not always necessary. The different classes of airspace, with their different flight rules, impacted on the risk management and conduct of the operation, making each operation different. The application of a "death mode fail safe" was required but was also highlighted as a complex process (Herrmann, 2016).

The world can also take their knowledge forward. For instance, they found that high-visibility paint and strobes make the aircraft more visible. They have also demonstrated that a RPAS operation can carry and utilize transponders and ADS-B, and utilize VHF airbands to help with being seen (Herrmann, 2016).

Incidentally, in 2007, an Aerosonde was the first unmanned aircraft to penetrate the eye of a hurricane.

Aerosonde are not alone in this extension of the boundaries, with many operators making inroads. Unfortunately, many of these inroads are still bounded by military or national security controls and are therefore not widely reportable.

Drone advertising

Marketers have a reputation for creatively applying new technology to the art of advertising. Drones are, not unexpectedly, appearing in advertising in abundance. But the creative minds behind advertising are also taking a fascinating range of other approaches to the public.

The shark-spotting drone under trial with the Australian Surf Rescue group took advantage of the media coverage and had the sponsoring bank's logo on the paintwork for the media launch (Nicholls, 2016).

Meanwhile, banner towing has been a longstanding advertising technique. There is nothing to stop drones doing the same thing within their given payload limits and the equipment is readily accessible for existing RPAS (Green Drone Advertising, 2016; Motor City, 2016). And then there's Russia, where an Asian restaurant employed drones for marketing purposes (Lee, 2014). These drones, which were equipped with advertisement flyers, were flown around the city of Moscow and displayed their ads to office workers just before lunch hour. If rules for very small drones and social acceptance of risks continue to mature, this could become a highly utilized, highly targetable, advertising technique.

Another transitional application is skywriting—when conducted by more nimble drones, the writing can be smaller, tighter and therefore closer and more personalized. No longer a "city-wide" event, marriage proposals could now be delivered directly to the family picnic.

Emergency!

During the June 2014 California wildfires, hot and dry weather conditions were hampering firefighting efforts. Eight new large fires were reported, bringing the total to ten with just one contained. The majority of the fire activity in California was centered around Los Angeles.

Four MAFFS C-130 air tankers and support personnel from the 146th Airlift Wing (California Air National Guard) and the 302nd Airlift Wing (Colorado Springs, Air Force Reserve) were already deployed to McClellan, CA to support wildland fire suppression operations.

The US Navy provided reconnaissance of wildfires in the rugged coastal mountains of California. Over four hours, they flew over multiple wildfires to assess wildfire progress through inaccessible terrain. Feedback to the Incident Management team was provided via live links, with possible mission redirection or refinement during flight (Naval Air Systems Command, 2012).

The US Navy's RQ-4A Global Hawk air vehicle can fly up to 60,000ft persistently for more than 30 hours, and can continue to feed imagery and other data obtained by the aircraft by satellite into the Navy ground segment.

Beyond the direct experience, there are a number of drone manufacturers that have developed water bombing-capable UAVs that are now being tested. Integration within the airspace between such aircraft and any manned aircraft will be a challenge, but perhaps moving to *all* unmanned aircraft in a full firefighting effort would be safer.

Crop spraying . . . dull, dirty and dangerous

Crop spraying is the practice of using an aircraft to apply fertilizer, insecticide or weed controls to a large acreage area. It is renowned in the aviation industry for being a dirty and dangerous job. Dirty, with constant chemical handling, and dangerous with low-level flying and constant maneuvers. Crop-spraying pilots are at constant risk of crash and of flying into structures and powerlines— for some it is considered dull, for others the risks bring a thrill to the job.

Most significantly, traditional aerial crop spraying incurs an average of one death per 100,000 hours flown (Struttmann and Marsh, 2004).

RPAS crop spraying has removed the pilot, and therefore the risk of death, from the activity.

Little wonder that the uptake of drone-based crop spraying has been so rapid. There are limits at this stage to the balance of cost efficiency between high-volume/high-acreage aerial crop sprayers, and the low-volume/low-acreage drone sprayers. However, investment continues worldwide in increasing the capability of drone crop spraying, and it is highly likely that traditional pilot-led methods of crop spraying will be virtually superseded over time.

Scanning and 3D mapping

Scanning and mapping the earth are no new developments. Mapping of the soil density, composition, reliefs, water concentrations and other features is common practice within industries such as mining, mapping, archeology and construction.

However, with the rise of UAVs, and the fitting of that scanning equipment to aircraft, the world can now be micro-scanned on a grand scale. A UAV can be equipped and pre-programmed to map an acre per hour, and can slog on for days (with stops for battery changes and a bit of a dust off).

Dr Catherine Ball is a high-flying scientist behind some world-leading, scanning-based, drone-supported research. Catherine uses drones to explore remote islands, map vegetation and help improve environmental conservation on engineering projects:

> In scientific projects and research work, being able to have an overall overview is of critical importance. It can likewise be helpful to see objects from different perspectives. At the same time it is very important not to disturb or influence the environment in which one is working.
>
> While the environmental impact of vehicles such as cars and helicopters can be considerable, thanks to breakthrough engineering, microdrones are perfectly suited to support projects and research work without disturbing the environment.
>
> (Ball, 2016)

A fabulous example, that is bound to make findings we will see in the news, is the work of scientists who are scanning the Amazon forest in Brazil (Amos, 2015). For this research, a drone is being sent up with a laser instrument to peer through the canopy for earthworks that were constructed thousands of years ago. The UK-led project is trying to determine how big these communities were, and to what degree they altered the landscape.

The toy conundrum

It is unfortunate, but the simple term "drone" can be quite misleading "since it implies that a two pound foam toy is equivalent to a large multimillion dollar military weapon system" (USA-OK, 2016).

As outlined in Chapter 2, the difference between an RPA that is a "model aircraft or toy" and the RPA that is treated by regulators (generally) is the *use* of the aircraft. In short form:

> Remotely piloted aircraft being used for profit are not toys.

However, this leaves a few gaping issues that regulators still grapple with—knowledge of responsibility. Where the flying of even the "toy" aircraft has some rules and regulations to comply with, to ensure everyone is safe regulators are taking a range of actions to educate users about the rules and their responsibilities. It is an uphill battle for them—with rates of sale escalating, and different uses compounding the issue.

Competition also drives change and escalation in the market, leading to the toy industry creatively inventing new toys that test even the fundamental principles being used to educate.

Take the flying fairy dolls that came out for Christmas 2013. They would fly across the room and dance above the child's hand. But they were capable of flying well above the house if released outside (YouTube, 2014) and were known to fly away.

Should regulators require educational information on "dolls boxes"?

Then there's the new Lego™-based kit from Flybrix (Flybrix, 2016). Flybrix provide a "Make Your Own LEGO® Drone Kit" to build the drones in any manner you can imagine—then fly them.

> Flybrix is a kit that can't be outgrown. Even experts can push their limits by tinkering with our powerful, programmable, custom PCB that's also Arduino compatible.
>
> (2016)

The kit includes their specialist software. This software could be utilized to ensure that whatever is built complies with the operational rules and doesn't head skyward into the airspace at Heathrow airport (see the UAV Outback Challenge in this chapter). But should it be regulated? Should purchasers be educated? Further, it's separable—we all know it's going to end up in pieces in children's toy boxes and building kits—so should instructions be embedded on the core motor? Furthermore, an unmanned aircraft can be built with many different construction toys and their robotic enhancements.

Then again, perhaps these toys are like remote control cars—there will always be kids who will drive those on the road, and the cars grown ups drive just run them over if they're in the way. A "life lesson" moment.

Moving on to more serious uses for drones, the medical industry is also taking advantage of the capabilities of these flying opportunities.

Fighting the Zika virus

In early 2016, Brazil was facing the worst-case scenario of Zika along with all other South American countries.

In the face of that scary scenario, Brazil used drones to help locate the breeding grounds of mosquitoes. Mosquitoes can carry a range of viruses, including malaria, dengue fever and the Chikungunya virus.

Sharma (2016) reports that the drones were deployed to locate and destroy breeding grounds for the mosquito, as they monitored the locations of breeding grounds with high accuracy. In Sao Paulo and several other towns, they made low-altitude rounds to detect the possibility of pests in gardens, on terraces and other breeding places. The operation inspected up to 60 million homes.

Once a target was detected, the drone was also equipped to fumigate the detected target right away.

> The idea of using the drones was born out of necessity. Many households were difficult to get into – some did not allow inspectors in, others simply had no one inside at the time. Drones greatly improve access without having to disturb residents.
>
> (RT News, 2016)

Robotic medical couriers

Way back in 2008 (aeons ago in drone industry terms), Professor Barry Mendelow and his team at the National Health Laboratory Service (NHLS) believed that UAVs could help save lives in South Africa, a country where access to medical services is highly varied.

For more than a decade, the team tested and researched what was then known as the "e-Juba"—iJuba is Zulu for pigeon, so e-Juba is the electronic pigeon.

> The lightweight electronic carrier pigeon can help save lives in some of the most remote parts of South Africa.
>
> (Bega, 2015)

The research was highly successful, and clearly showed how drone medical couriers can ferry medical cargoes between rural clinics and urban centers. The trials, comprised of 300 fully audited flights of up to 30km, were concluded in 2010, and proved how reliable the technology was. During over 300 test flights the team didn't lose a single cargo item or artifact. The trials also demonstrated the potential for dramatic savings in costs compared with their existing land-based transport options (Bega, 2015).

According to Bega (2015), Mendelow is very clear about the value of utilizing drones. "When we started our project . . . the average wait was six weeks for a diagnosis of tuberculosis. Now it can be done with this mobility in one day." South Africa, Mendelow points out, has the highest annual incidence of tuberculosis in the world, with the "poor logistical support for the diagnostic process" cited as one of the major bottlenecks in the effective treatment of rural patients.

> Another purpose was to transport much-needed therapeutic supplies such as blood, snakebite serum or rabies immune globulin, from central depots such as hospitals to remote clinics.
>
> (Bega, 2016)

In 2009, the NHLS approached S-Plane Automation to create a small, inexpensive UAS to transport sterile medical samples between more than 1,500 rural clinics and laboratories.

Mendelow said:

> The resulting system, named Nightingale, is an incredibly reliable aircraft, capable of enduring extreme punishment in remote parts of South Africa. Nightingale was a resounding success. It completed a two-month-equivalent operational field trial between a clinic and a laboratory on the West Coast of South Africa with a 100% success rate under extreme environmental conditions.
>
> (Govender, 2015)

S-Plane's Nightingale UAS is a highly reliable, state-of-the-art autonomous Mini-UAV specifically designed for transport and airdrop of medical samples of up to 150g. The UAV was developed in conjunction with the NHLS of South Africa to transport sterile medical specimens from remote rural clinics to pathology laboratories. "An autonomously controlled payload release bay on the under-side of the aircraft allows samples to be airdropped precisely at specified GPS coordinates" (Govender, 2015).

Okafor (2008) notes however the perpetual issue: "For this to become a practical enterprise, however, demands more than just having the requisite technology, as formidable non-technical barriers remain. These include solving the legislative and financial issues."

What's happening in those BRIC countries?

For many UAV companies, including American ones, the BRIC countries (Brazil, Russia, India and China) are appealing sites for research, development and manufacture. So they are worth a quick review to see what's going on.

Russia

With regard to Russia's UAV industry, with the USA intervening and Russian manufacturers focused on equipping UAV, Russian focus on UAVs seems to be tied to military (real drone) uses and commentary. Watch this country for military drone technology development, and its subsequent cascade down into commercial UAVs.

India

For India, the UAV industry seems to be coming into focus, with start ups gaining foreign attention and investment in research. They aren't leaders at this stage, and don't yet seem to be engaging with plans to lead (publicly). Watch this country for novel and creative UAV applications and add-ons.

China

UAVs are nicknamed "aerial robots" in China. Dronelife (2014) reference sources from the UAV industry in China estimating that "given the trend for unmanned, compact and intelligent aerial equipment, UAV demand in China could reach 46 billion yuan (US$7.4 billion) over the next two decades."

China Securities Journal (Dronelife, 2014) believes that although there is a lot of potential for UAVs in the Chinese market, there are still several challenges for aircraft manufacturers, including meeting industrial standards and acquiring aviation permits.

According to Keck (2014) a state-owned Chinese defense company will be the largest UAV manufacturer (worldwide) over the next decade, based on data from Forecast International, a private market researcher. "The report predicts that the global drone market will more than double in the next ten years, rising from $942 million in 2014 to an annual $2.3 billion in 2023."

Keck (2014) states that the report forecasts that the Aviation Industry Corporation of China (AVIC), a state-owned Chinese defense company, will lead the world in UAV production: "According to Forecast International, AVIC will produce about $5.76 billion worth of UAVs through 2023. This is more than half of the UAVs by value that will be produced during this time period. Nearly all these will be sold to Chinese consumers." The AVIC and its subsidiaries already produce a number of UAVs for the Chinese market.

China, in particular AVIC, is probably going to have significant control over the prices and economic expansion of UAVs in the coming years.

Brazil

In June 2014, Brazil became the first Latin American country to export home-grown UAVs (to two undisclosed African countries) and is attracting foreign technology and investment in UAVs and systems.

Pool (2015) quotes Alejandro Sánchez (Research Fellow at The Council on Hemispheric Affairs) in reporting on the Brazilian drone industry. According to Pool (2015), Sánchez says:

> one reason for Brazil's rapid ascension in the drone revolution can be found in the history of the military dictatorship. Military rulers built on Brazil's already formidable industrial prowess by nationalizing key sectors and investing significant state resources toward the development of a military industrial complex. Thus, the dictatorship of Brazil stood out from contemporaries in Chile and Argentina, by cultivating an international reputation as an exporter of quality weapons and aircraft.

Embraer (Brazil's state-owned aeronautics company), established in the 1980s, is currently the fourth largest aircraft manufacturer in the world, after Airbus (EU), Boeing (USA) and Bombardier (Canada).

Brazil and its counterparts also offer a highly unregulated airspace to companies fleeing the strict regulations of the American FAA. Brazil's deregulated airspace and the absence of a rigorous permit system, significantly lowers research and development costs for foreign UAV producers (Pool, 2015).

Investments by foreign manufacturers will probably transform Brazil into a regional base of drone development, and perhaps this will lead to a strong position for them producing drones for the world market.

Bio-inspired UAS

UAS autonomy is held out as the next big change for the technology. For true autonomy, as described earlier in Chapter 5, there is a long way to go. However, the core need for UAVs to highly effectively detect and avoid obstacles and areas is perhaps closer to being achieved. A fascinating approach to finding the solution is bio-robotics.

> System autonomy . . . focuses on the development of advanced guidance, navigation, sensing, communication, and safety systems for unmanned aircraft. Increasing autonomy reduces the need for high bandwidth communications links, decreases pilot workload, and can lead to improvements in safety and mission performance.
>
> (RMIT, 2016)

RMIT in Australia (O'Malley, 2016) are researching:

> the design and analysis of bio-inspired systems for UAS. Research includes the study of natural flyers (e.g. birds and insects) and their natural environment (flows and turbulence) for innovations in the design of micro air vehicles.

In short hand, this is the design of even more "bird-like" air vehicles, with more natural wing designs and flapping motions.

Meanwhile, researchers from the University of Illinois (Ahlberg and Neilson, 2016) have produced a new generation of muscle-powered biological robots, or "bio-bots" that can be stimulated to walk using electrical impulses (they walk on their own, powered by beating heart cells from rats). They refer to this integration of nature-based programing into robotics as "soft robotics."

The researchers emphasize that the marriage of soft robotics with living biological components, such as cells and tissues, permits the development of machines that are able to sense and respond to a variety of controlled environmental stimuli.

MIT takes a different tack (Conner-Simons, 2016), being focused on the software design being biological mimics—and developed motion-planning algorithms that allow drones to do donuts and figure-eights in object-filled environments. MIT researchers have developed UAV that can operate autonomously in constrained spaces and unmapped environments.

Now, those three research teams need to come together—imagine a UAV with self-propelling biological wings and obstacle sensing!

The UAV outback challenge

Given the wide expanses of space available, and rather thin aircraft coverage, it's not surprising to find that Australia has a unique research program underway for UAVs. The self-stated goal of the UAV Challenge is to:

> demonstrate the utility of Unmanned Airborne Vehicles (UAVs) for civilian applications, particularly in those applications that will save the lives of people in the future . . . [by] harnessing the ingenuity and passion of aero modellers, university students and high-school students around the world to develop novel and cost-effective solutions.
>
> (UAV Challenge, 2016)

The challenge has been the subject of a few documentaries, but a wonderful feature on the 2014 Search and Rescue Challenge by BBC News is worth checking out if you are interested in more (six minutes long, www.bbc.co.uk/programmes/p02crqy5, accessed April 17, 2016).

The challenge is to complete a set mission—where that mission revolved around a search and rescue-type scenario where an aircraft would have to be used to find a lost bushwalker, known as Outback Joe (played by a friendly local mannequin in high-visibility gear), and then drop a water bottle to him. Outback Joe was left in a search area measuring 1.26 km × 2.26 km with the closest point of that search area 3.4km from the end of the runway at the airport take-off location.

> The UAV Challenge Outback Rescue was originally developed to promote UAV's significance to Australia. The UAV Challenge has been a joint initiative between Queensland University and Technology – QUT (through its Australian Research Centre for Aerospace Automation) and CSIRO. CSIRO and QUT have been joined by other partners during the years including the Queensland Government, Boeing, Aviation Development Australia Limited and AUVS-Australia.
>
> (Roberts *et al.*, 2015)

The challenge opened in 2007, and was run regularly until it was achieved (Joe finally got some water, phew!) in 2014. Check out how the challenge was finally won at Roberts *et al.*, 2015.

To keep up the pressure to innovate, a new challenge has been set, focused on pushing the skills, learning and development further. To discover the new challenge head to https://uavchallenge.org/ (as at August 14, 2016).

Why is this challenge so important?

The UAV Challenge event achieved far more than we as the organisers dared to dream. Some of this was due to the longevity of the competition, in that it is more likely that positive outcomes will be achieved over a long period of time. The number of people exposed to the field of civilian UAVs due to the competition was large. As well as the over 2000 team members that took part in the Search and Rescue competitions and the over 400 high-school students that took part in the Airborne Delivery Challenge event co-located with the main event, millions of members of the general public were exposed to [the] possibility of UAVs being capable of saving lives in the future . . . .

(Roberts *et al.*, 2015)

The first example of a commercial system developed for use in the UAV Challenge was the Millswood Failsafe Device. This was an electronic device designed to supplement commercial autopilots . . . to implement all the failsafe requirements of the UAV Challenge Search and Rescue mission. A manufacturer had recognised that there was a gap in the market and had created a new product . . . Significantly, the manufacturer reported that they had sold this device worldwide to others who realised the need for safer unmanned aircraft operations, including auto pilot developers who require backup systems when developing and testing brand new products.

(Roberts *et al.*, 2015)

The key driver for ensuring that each competing aircraft had a failsafe capability was that it was clear that reliable communications between a team's ground control station and the aircraft was difficult. The UAV Challenge search and rescue mission required reliable radio communications over at least a 7km range. Most teams struggled with the communication issue in the early years of the UAV Challenge. A company called RF Design set about using some existing low-power chipsets with an integrated sub-GHz RF transceiver as the basis of their new data modem, resulting in the RFD900 data modem. By 2014, all but one of the teams that flew at event was using the RFD900, including the winning team. RF Design report that they have sold several thousand RFD900 modems worldwide, mainly in the UAV sector. They are now widely used by UAV professionals and have been tested to an altitude of 60,000ft and to ranges up to 67km line-of-sight.

(Roberts *et al.*, 2015)

When the UAV Challenge was announced in 2006, teams had two choices when it came to civilian autopilots. They could spend thousands of dollars and pick between a limited number of high-end products, or they could develop their own from scratch. In 2014 the situation was very different with multiple affordable autopilots readily available to all teams with the UAV Challenge playing its part in motivating these developments. Some of the teams from 2009 began using the open source Paparazzi Project autopilot (software and hardware) and the rules of the UAV Challenge began to influence features being developed for that autopilot. In 2010, the Ardupilot and PX4 open source autopilot communities began to take notice of the UAV Challenge, and those autopilots began to adopt features specifically to address the competition rules. At the 2014 UAV Challenge event, core developers from all three of these major open source autopilot communities were members of competing teams. By 2014 much of the standard flight termination safety code for all three autopilots was written specifically to comply with the UAV Challenge rules.

(Roberts *et al.*, 2015)

So you see, testing and playing with UAVs has developed technology and software that improve the safety of UAV operations. Nice.

Organized experimentation

However, from there, it contributes even more. For Australia, there is now a strong student base for UAV skills—who have cut their teeth on the technology and are coming out of the universities already, while more flow into the education system as time goes on. So it's not just pilots who are being inspired and home grown, but also roboticists, manufacturers, software programmers and so on, which could see them in a world-leading position for this industry in the years to come.

Australia isn't alone in establishing testing protocols and sites. MIT in the USA (Litant, 2014) have worked with a consortium of interested parties to find, propose and now establish a realistic UAS test site in Massachusetts. While not a unique proposition, the site is also paired with a site at Griffiss International Airport in Rome, NY. Together, they establish a structured, realistic, cross-country test environment for RPAs (Litant, 2014).

The test sites have already made remarkable gains, such as night-time operations, flying multiple aircraft in the same airspace and researching and testing aircraft up to 1,200 feet. Nevertheless, there's much left to do, and that will require investment and support from industry partners. Those partners will be much more likely to use the FAA test sites if they can be sure those sites will be operational beyond the end of Fiscal Year 2017.

(Nuair, 2016)

They have also been testing what is known as an "optionally piloted" aircraft; that is, one where the pilot is not only not onboard, but doesn't have to be in control of the aircraft (and can walk away (for a quick coffee only, surely?)).

But let's move on to a more pressing and immediate issue—pizza delivery.

As the world of UAVs became more than just a geek hobby, as everyone started to jump on the bandwagon—lots of companies came out with ideas to deliver products to you via UAV. Ideas range from pizzas, books and parcels to medical items and legal documents. And yet, years into this revolution, I still have to pick my pizzas up (sad face).

What is going on?

In truth, there is quite a lot that has to be sorted out in order for personal home delivery to become reality. Issues include being able to fly over/near people (or more to the point, fly so that you are not near them), keeping the package safe/hot/dry, placing/handing over the package, whether you can fly "as the crow flies" (over houses) or have to take a specified route (and which one that is), RPAS flight time (an hour or two is fairly limited) and the security of the RPAS itself (from eagles, powerlines and random acts of theft/ interference).

The concept of home delivery by UAV is still in the pipeline, but the reality of the matter is that it is still in testing and proof of concepts phases.

Australia's Postal Service (Australia Post) has successfully tested drones for delivering small packages (Ferrier, 2016). Australia Post received the backing of the country's CASA to deliver packages using drones in a limited test across 50 locations. A customer trial is in planning now (Cuthbertson, 2016).

> This closed-field trial is an important next step in testing the new technology which will potentially deliver small parcels safely and securely to customers' home, allowing for faster transportation of time-critical items like medication.
>
> (Australia Post, 2016)

Australia Post envisages that the drones would be used for time-critical deliveries or for where there are significant distances between the road and the front door (Australia Post, 2016), however they emphasize that regular drone delivery may be up to ten years away (Cuthbertson 2016; Ferrier, 2016).

Amazon Prime Air is the Amazon business that is:

> a future delivery system from Amazon designed to safely get packages to customers in 30 minutes or less using small unmanned aerial vehicles.
>
> (Amazon, 2016)

Amazon are currently testing multiple vehicle designs and delivery from their "Prime Air" development centers in the US, the UK and Israel. Amazon has said it would be ready to begin delivering packages to USA customers via drones as soon as federal rules allow (Layne, 2015). They haven't yet made a commitment to when their deliveries will start in other countries but they do note that their deployments are dependent on having "the regulatory support needed to safely realize our vision" (Amazon, 2016).

Google publicly launched their efforts to deliver packages via drone in 2014, and still seem to plan to launch the delivery service in 2017. Their project is known as Project Wing. According to the patent, which was filed back in 2014 but only granted now, Google has developed "mobile delivery receptacles" that work hand in hand with its delivery drones. Effectively, the mobile delivery receptacles are remote boxes on the ground with wheels. They communicate and guide the drones in the sky via infrared beacons or lasers. Once located, the drone flies down to ground level and transfers its package into the mobile delivery receptacle, which then secures it and scurries off to a secure holding location (Grothaus, 2016).

Using this robotic communication system, the drone will eventually lower the package down to the robot. The delivery robot will then transport the package to a secure location, "such as into a garage," the patent reads (Muoio, 2016).

Walmart Stores Inc. have also been working to test drones for home delivery, curbside pickup and checking warehouse inventories. Layne (2015) considers this a sign it plans to go head-to-head with Amazon in using drones to fill and deliver online orders. The world giant retailer has been conducting indoor tests of small UAS and in 2015 sought for the first time to test the machines outdoors.

In addition to having drones take inventory of trailers outside its warehouses and perform other tasks aimed at making its distribution system more efficient, Walmart is asking the FAA for permission to research drone use in "deliveries to customers at Walmart facilities, as well as to consumer homes," according to a copy of the application (Layne, 2015).

The retailer also wants to test drones for its grocery pickup service, which continues to expand. The test flights would confirm whether a drone could deliver a package to a pickup point in the parking lot of a store, the application says (Layne, 2015).

Walmart also said it wants to test home delivery in small residential neighborhoods after obtaining permission from those living in the flight path. The test would see if a drone could be deployed from a truck "to safely deliver a package at a home and then return safely to the same." (Layne, 2015).

At this point in time it is down to a race across the world. Skype, Maersk, Alibaba and many others are testing and trialing drone deliveries. There is immense pressure in the USA, with the FAA implementing a drone registration system to support the concepts, so the UAS may well come in first for a permanent successful service, assuming their Federal Administrators can develop appropriate rules and release them in a reasonable timeframe.

That still doesn't deliver my pizza via drone all hot, fresh and cheesy for me, in the back blocks of Australia.

Practical applications

I have saved my favorite until last—RoboBees.

Coming out of Harvard, and inspired by the biology of the bee and their hive behavior, this is a swarm of co-dependent UAVs. "The RoboBee is so small and lightweight that it cannot break the surface tension of the water." (Burrows, 2015). The practical applications of the RoboBee are (Wyss Institute, 2016):

- autonomously pollinating a field of crops;
- search and rescue (e.g. in the aftermath of a natural disaster);
- hazardous environment exploration;
- military surveillance;
- high-resolution weather and climate mapping; and
- traffic monitoring.

These are the ubiquitous applications typically invoked in the development of autonomous robots. However, in mimicking the physical and behavioral robustness of insect groups by coordinating large numbers of small, agile robots, Harvard researchers believe they will be able to accomplish such tasks faster, more reliably and more efficiently (Burrows, 2015).

More broadly, the scientists anticipate the devices will open up a wide range of behavioral understanding discoveries and practical innovations, advancing fields ranging from entomology and developmental biology to amorphous computing and electrical engineering.

Through a relationship with the Museum of Science, Boston, the team will also create an interactive exhibit to teach and inspire future scientists and engineers . . . I am booking tickets as soon as that exhibit is in place!

Brain

One of the most complicated areas of exploration the scientists will undertake will be the creation of a suite of artificial "smart" sensors, akin to a bee's eyes and antennae. Professor Wei explains that the ultimate aim is to design dynamic hardware and software that serves as the device's "brain," controlling and monitoring flight, sensing objects such as fellow devices and other objects, and coordinating simple decision-making (Wyss Institute, 2016).

Colony

Finally, to mimic the sophisticated behavior of a real colony of insects will involve the development of sophisticated coordination algorithms, commun-ications methods (i.e. the ability for individual machines to "talk" to one another and the hive) and global-to-local programming tools to simulate the ways groups of real bees rely upon one another to scout, forage and plan (Wyss Institute, 2016).

Burrows (2015) reports progress:

> "Through various theoretical, computational and experimental studies, we found that the mechanics of flapping propulsion are actually very similar in air and in water," said Kevin Chen, a graduate student in the Harvard Microrobotics Lab at SEAS. "In both cases, the wing is moving back and forth. The only difference is the speed at which the wing flaps."

Coming from the Harvard Microrobotics Lab, this discovery can only mean one thing: swimming RoboBees. The RoboBees can already transition from flight to the water to swim, but research is ongoing to develop techniques to break from the water back into flight. You can follow this work at http://robobees.seas.harvard.edu/ (as at May 15, 2016).

References

Ahleberg, Liz and Nielsen, Josh "Building with Biology: Tiny Bio-bots Powered by Living Cells", http://engineering.illinois.edu/do-the-impossible/biobots/, accessed August 14, 2016.

Aerosonde www.aerosonde.com/, 2015, accessed April 26, 2016.

Amazon www.amazon.com/b?node=8037720011, 2016, accessed April 20, 2016.

Amos, Jonathan "Brazil Amazon: Drone to Scan for Ancient Amazonia" BBC Science and Environment, February 14, 2015, www.bbc.com/news/science-environment-31467619, accessed April 20, 2016.

Australia Post "Australia Post Delivery Trial Takes Flight" Auspost Newsroom, https://auspost.newsroom.com.au/Content/Home/02-Home/Article/Australia-Post-Delivery-Trial-Takes-Flight-/-2/-2/6092, April 15, 2016, accessed April 20, 2016.

Ball, C "Microdrones" www.microdrones.com/en/applications/areas-of-application/science-and-research/, accessed April 20, 2016.

Bega, Sheree "These Drones Bring Healing, Not War" National Health Laboratory Service, NHLS News, October 8, 2015, www.nhls.ac.za/?page=news&id=4&rid=600, accessed April 14, 2016.

Burrows, Leah "Dive of the Robobee: Harvard Microrobotics Lab Develops First Insect-Size Robot Capable of Flying and Swimming" October 21, 2015, Harvard John A. Paulson School of Engineering and Applied Sciences News and Events, www.seas.harvard.edu/news/2015/10/dive-of-robobee, accessed August 14, 2016.

Conner-Simons, Adam "Drones Dodge Obstacles – Motion-Planning Algorithms Allow Drones to Do Donuts, Figure Eights in Object Filled Environments" MIT News, January 19, 2016, http://news.mit.edu/2016/csail-drones-do-donuts-figure-eights-around-obstacles-0119, June 10, 2016.

Cuthbertson, Anthony "Australia Post to Launch Drone Delivery Service" *Newsweek*, April 19, 2016.

Dronelife "Chinese Companies Making Inroads Into UAV Market" Dronelife.com, July 15, 2014, http://dronelife.com/2014/07/15/chinese-companies-making-inroads-uav-market/, accessed April 20, 2016.

Ferrier, Stephanie "Australia Post to Trial Drone Parcel Delivery of Online Shopping" ABC News, www.abc.net.au/news/2016-04-15/australia-post-to-trial-drone-parcel-delivery-of-online-shopping/7331170, June 10, 2016, accessed April 15, 2016.

Flybrix https://flybrix.com/, accessed April 20, 2016.

Govender, Kemantha "Drones in Healthcare" University of the Witwatersrand, Johannesburg, General News, www.wits.ac.za/news/latest-news/general-news/2015/2015-10/drones-in-healthcare.html, October 6, 2015, accessed April 20, 2016.

Green Drone Advertising, http://greendroneads.com/portfolio/next-level-banners/, accessed May 20, 2016.

Grothaus, Michael "This is How Google's Project Wing Drone Delivery Service Could Work" www.fastcompany.com/3055961/fast-feed/this-is-how-googles-project-wing-drone-delivery-service-could-work, January 27, 2016, accessed April 20, 2016.

Herrmann, Paul "Challenges & Opportunities for Civil Airspace Integration of Small UAS in Australia, Textron Systems" Aerosonde, Pty Ltd, Textron Systems Paper, Australian Association for Unmanned Systems, RPAS in Australian Skies Conference, March 2016.

Keck, Zachary "China to Lead World in Drone Production" *The Diplomat*, May 2, 2014, http://thediplomat.com/2014/05/china-to-lead-world-in-drone-production/, accessed August 14, 2016.

Layne, Nathan "Exclusive – Wal-Mart Seeks to Test Drones for Home Delivery, Pickup" Reuters UK, http://uk.reuters.com/article/uk-wal-mart-stores-drones-exclusive-idUKKCN0SK2JA20151026, October 26, 2015, accessed June 10, 2016.

Lee, Joel "5 Amazing Uses for Drones in the Future" MakeUseOf, www.makeuseof.com/tag/5-amazing-uses-drones-future/, November 24, 2014, accessed April 20, 2016.

Litant, William "Cape Cod Drone Test Site Will be Boon to MIT Researchers" MIT News, http://news.mit.edu/2014/cape-cod-drone-test-site-will-be-boon-to-mit-researchers, January 17, 2014, accessed April 20, 2016.

Motor City http://motorcitydroneco.com/products/professional-banner-rig, accessed April 20, 2016.

Muoio, D "Google is Thinking about Using Robots on Wheels to Deliver Packages" *Tech Insider*, January 29, 2016.

Naval Air Systems Command "Navy UAV Crashes" Story Number: NNS120611–09, www.navy.mil/submit/display.asp?story_id=67739, November 6, 2012, accessed April 25, 2016.

Nicholls, Sean "Shark Spotting Drone the 'Future of Rescue' in NSW" *Sydney Morning Herald*, www.smh.com.au/nsw/shark spotting-drone-the-future-of-rescue-in-nsw-20160228-gn5leq.html, February 28, 2016, accessed April 17, 2016.

Nuair http://nuairalliance.org/capabilities/uas-traffic-management/, Facebook post April 19, 2016, accessed April 20, 2016.

Okafor, Emeka "E-Juba: Medic Air Couriers" *Timbuktu Chronicles*, http://timbuktu chronicles.blogspot.com.au/2008/09/e-juba-medic-air-couriers.html, September 9, 2008, accessed April 20, 2016.

O'Malley Sean "Researchers Send High-flying Drones Soaring" *RMIT University News*, www.rmit.edu.au/news/all-news/2016/april/research-sends-high-flying-drones-soaring/, April 11, 2016, accessed June 10, 2016.

Pool, Tim "The Scary History and Future of Brazil's Booming Drone Market" *Fusions News*, http://fusion.net/story/187490/brazil-drone-laad-conference/, August 24, 2015, accessed April 20, 2016.

RMIT "Unmanned Aircraft Systems" http://rmit-test.adobecqms.net/research/research-institutes-centres-and-groups/research-institutes/platform-technologies-research-institute/programs/innovative-engineering-systems-program/unmanned-aircraft-systems-uas, accessed June 10, 2016.

Roberts, Jonathan, Frousheger, Dannis, Williams, Brendan, Campbell, Duncan and Walker, Rod "How the Outback Challenge Was Won: The Motivation for the UAV Challenge Outback Rescue, the Competition Mission, and a Summary of the Six Events" *IEEE Xplore*, Volume PP, Issue 9, April 13, 2016, http://ieeexplore.ieee.org/xpl/login.jsp?tp=&arnumber=7452338&url=http%3A%2F%2Fieeexplore.ieee.org%2Fxpls%2Fabs_all.jsp%3Farnumber%3D745233, accessed August 17, 2016.

RT News "Brazil using Drones to Fight Zika, Cuba Deploys Army" www.rt.com/news/333264-brazil-zika-drones-cuba/, February 22, 2016, accessed April 20, 2016.

Sharma, Trivesh "Brazil Deploys Drones to Fight Zika Virus" Australia Network, www.australianetworknews.com/brazil-deploys-drones-fight-zika-virus/, February 2016, accessed April 14, 2016.

Struttmann, TW and Marsh, SM "Work-Related Pilot Fatalities in Agriculture – United States, 1992—2001" (2004) 53(15) *Centre for Disease Control and Prevention, Morbidity and Mortality Weekly Report* 318–320, www.cdc.gov/mmwr/preview/mmwrhtml/mm5315a4.htm, accessed April 20, 2016.

UAV Challenge "About" https://uavchallenge.org/about/, accessed June 10, 2016.

USA-OK (Unmanned Systems Alliance of Oklahoma). "Myth vs. Facts about Unmanned Aircraft ('drones')", www.ocgov.net/oneida/sites/default/files/userfiles/airport/UASHobbyistRefInfo/Myths%20vs%20Fact.pdf, accessed August 14, 2016.

Wyss Institute "Autonomous Flying Microrobots (RoboBees)" http://wyss.harvard.edu/viewpage/457, 2016, accessed April 20, 2016.

YouTube "These Girls Underestimate the Power of Their Flying Fairy Doll", www.youtube.com/watch?v=m8tS0W3NIfA, October 4, 2014, accessed April 20, 2016.

10 Where no one has gone before
The future with drones

Let us put our minds together and see what life we can make for our children.
Sitting Bull

Where to from here?

Futurologists approach this question scientifically, with a strong body of trends and evidence behind them. Science fiction writers stretch imagination beyond the provable into the truly creative space.

Given the current rate of technology development, the likelihood of disruptive change and low levels of data in the drone space—this chapter leans more toward science fiction than futurology.

Data and current predictions

Let us start with the facts—Teal Group studies (Finnegan, 2015) predict a doubling in the value of the UAV market. This includes jobs, products, attached equipment and related industries. UAVs are, without question, becoming more predominant. You can expect to see them more often, doing more things, and you will know more people who own or use one.

There is very little data to predict the change and impact of drones. Teal Group (Finnegan, 2015) are predicting (over a decade) up to 30 percent growth in the industry, with a doubling in spending on drones, tripling in the value of manufacturing, an overall market value near a US$100 billion and a market for drone payloads in excess of US$6 billion.

In Australia, the growth in approved drone operators is currently exponential, although these figures are influenced by the rate of processing of applications by the regulator, so the trend is unreliable. Much of the world is still gearing up to approve drone operations; so many operations are unapproved (not illegally so) and also untraceable.

There are few to no figures in place yet for the number of manufacturers, types of product, range of uses or profit from use. Equally, there is no visibility yet of operators who have already dissolved or shifted markets.

One prediction I'd personally bet on—with a multibillion dollar industry coming on, data will start to emerge, and potentially become its own mini industry within data markets.

But what does growth mean? What impact will they have on our lives?

Cargo delivery?

The mobility of unmanned aircraft makes them an obvious choice for moving those items around that we seek to move. It is no secret that multiple corporate entities (Australia Post, Google, Amazon) are working to deliver parcels "by drone." True, they aren't delivering yet, as there are a few wrinkles to sort out, and some social acceptance to negotiate.

But take it as read that as those issues are addressed, parcels can be delivered. If a parcel can be delivered, it is a short stretch to have any form of cargo delivered. What cargo would you request and value having delivered cheaper, faster or maybe have it follow you?

•   Gourmet dinner from your favorite restaurant?
•   An organ for transplant, bypassing traffic jams?
•   Novel delivery of flowers and chocolate to your picnic spot?
•   Last-minute gift to mum?
•   Bring in a crate of snow from New Zealand?
•   Receive your new luxury car, direct from Europe, in just days?

Why not?

Stretch this concept a step further—given that UASs are so affordable, and the next generation will be "UAS" natives who can probably fly one just as easily as they ride a bike, it won't be long before every household has a UAS and every adult has a UAV controller's certificate. It'll be just as common as cars. Can an unmanned aircraft replace a car? Try switching the cargo concept around, from delivery to you—to delivering "from you."

•   Snowed in? (Or maybe it's 40° C and the car air-conditioner is broken?)— rather than having your groceries delivered, now you could order online and go and collect them for yourself. Easier when you can choose the when and where?
•   Grandma can't drive anymore? Sling a pod under the unmanned aircraft and maybe you can pick her up and bring her over for lunch "as the crow flies" (if she's up for it).
•   Drop off library books. Deposit banking and cheques. Deliver dinner to a sick friend. Provide original signed documents to your lawyer, or ex-spouse.

How will this ease of transporting items affect us? Even less reason to leave the house than ever—do more of our society end up as shut ins? Or do our socially limited society members become more enabled and empowered—with greater social participation? Do obesity issues increase because we do even less movement than ever? Or does it free us up to just walk and enjoy experiences?

Flying robots?

Take the thought experiment another step forward—unmanned aircraft are not only mobile, but highly flexible. Imagine the not too distant future when robotics is closely involved with your personal unmanned aircraft, that there comes time when you can buy affordable roboticized attachments for your unmanned aircraft. Not just cameras and cargo pods, but functional and mechanized attachments.

We already have robotic vacuum cleaners and lawn mowers. It won't be long before you can clean your gutters and windows with the one machine, replace the robotic lawnmower and clean out the spider webs. Wash the car roof (a definite appeal for the shorter members of society), paint the walls, trim the tree tops. . . .

Take that back into an industrial context—there are already unmanned aircraft that can harvest fruit on farms so why not reap the wheat, spray the crops, fertilize and turn the soil, and deliver the wheat to market all with the one machine? Or wash super structures, replace their window seals, fight a fire and service their air-conditioning units with one unmanned aircraft?

Does our physical and transport workforce retract significantly? Or do new opportunities for the skill sets open up? Do flying robots empower us to do less, or do we follow our current trend and use the new service to allow us to do even more than ever, faster?

Travel?

Take the thinking down a different path and test the idea that the unmanned aircraft is also autonomous, or safely automated. This takes you beyond just a remotely piloted vehicle, to one that has no dedicated pilot . . . that can navigate for itself. Perhaps to avoid the "random humans" these autonomous vehicles are all required to transit at least 10ft above us along the roadways. Perhaps they have pre-programmed rules of the air and detailed local maps in 3D.

Do you now prefer to pre-program your unmanned aircraft to take you to work in a climate-controlled pod installed underneath your unmanned aircraft? Where you can work or watch the world go by? Then send it home to take the children to school? Is this now driverless cars at the next level? The start of the Jetsons' era? It could return home to park, which would save parking expenses and, given that vehicles would be able to return to base cities globally, there would need only one space per vehicle, giving a new lease of life to multi-storey carparks. Could this double the density of city living as a result? Closer and easier access to facilities and services—but will there be increased or decreased fossil fuel use?

The ICAO is already working on the SARPs to facilitate unmanned international flights. Imagine perhaps the impact of remote piloting on large aircraft, that the Airbus and Boeing aircraft of the world are retrofitted to remove the pilot from the cabin? International travel becomes an unmanned exercise. International pilots can now work from anywhere in the world and live at home. Does the pilot's role pay less and make international travel more affordable? Does the removal of the cockpit allow a few extra seats and make travel just that bit more affordable again?

Or, given the flexibility of the pilots—do international flights become "bespoke," getting smaller and flying directly between cities? Instead of boarding a flight in Adelaide for Sydney, Sydney for Los Angeles and Los Angeles to New York, you might be onboard an aircraft that seats five people, with self-service food and drink, one toilet, room to recline, 360-degree views and flying direct Adelaide to New York—saving 15 hours of transit time each way.

Then again, there's a third option. Perhaps RPA could replace you on your travels? Attach a screen to the machine, and it can display your face, as you do now with video conferencing. The use of a RPA would allow you to not only attend a meeting internationally, but the mobility of a small RPA might allow you to fully participate in the conferences, evenings and events by moving from room to room, putting up a hand to ask a question, and developing networks remotely all while maintaining a personal connection.

The internet is already a great place for lounge suite travel, with RPA-generated footage from flights over iconic places. The next level, controlling your own flight in and around these icons is possible. Try the idea of renting a camera-enabled RPA in Machu Picchu and do your own tour, in your own time—catch a sunrise and sunset, stay overnight, hop across to Las Vegas the next day, and over to Florida for an afternoon. Congratulations, you have just traveled internationally on your weekend, without a passport, customs issues, jet lag, gastric disturbances or transit time. Take a refugee who can never travel home physically, or someone with medical restrictions that prevent flying, to iconic places in unsafe countries. For those facing travel limitations it might be their best travel option—and perhaps create a market in "semi-international travel."

It is conceivable that the long-term impact of RPAs may be to decimate the international travel industry. Certainly, there will always be those who will want the experience of travel, to warmer climes, or snowfields. But what if work-based international travel were to halve over a decade? Or tourist travel fell by the wayside?

Take another technology onboard though, marry a unmanned aircraft with supersonic flight, and perhaps the scenario is different again. A number of countries are working to develop supersonic aircraft to replace the Concord. No longer hampered by the compromise of shaping the aircraft (or in the case of the Concord, bending it) so the pilots can see for landing, the project is a new and open field. Sleeker, more streamlined aircraft. Would they be therefore more affordable, and make supersonic travel more reachable? London to Sydney would drop from 20-odd hours, to around eight hours. Suddenly the whole world is just a weekend trip for anyone, from any point on the planet to any other point. Does the world, and distance, shrink even further? Would this negate the market for subspace (hypersonic) travel? Would you think differently if you knew this form of travel also reduced greenhouse gas emissions?

Take a different approach though, how about travel from your hospital bed?

One innovative application of video games in health care is their use in pain management. The degree of attention needed to play such a game can distract the player from the sensation of pain, a strategy that has been reported and evaluated among paediatric patients. Controlled studies show that video games can provide cognitive distraction for children and reported that distracted patients had less nausea and lower systolic blood pressure than controls (who were simply asked to rest) after treatment and needed fewer analgesics (Griffiths, 2005).

Imagine instead of video games, giving an unmanned aircraft to adult patients. Would the focus required be as distracting as a video game for children? Would the ability to "leave the hospital bed" be beneficial for chronically ill patients? There is a lot to point to a high potential for patients, so don't be surprised if your health insurance one day includes access to a drone, and your local hospital grounds and nearby parks have "therapeutic" drones meandering around, "traveling from the bedside."

Emergency services?

The emergency services teams are an interesting mix—as both early adopters and late adopters of RPAS technology. Early adopters in that they are empowered in emergencies to call on military drones and use them wherever and however might be useful. Thus demonstrating the technology capabilities. But also late adopters, in that emergencies are no time to be working with untested technology and new processes, so the uptake of bespoke solutions is far slower than other industries.

There is strong public support for emergency services usage of RPAS, where tolerance of the technology is greater for "the greater good." How far this tolerance will extend is still being tested. It is not a far stretch to see a time when the RPA will arrive before an ambulance, having been able to go over the top of traffic in densely populated cities like New York, London and Beijing. They will probably also go to multi-storey buildings, entering through windows to access higher floors for patients. Imagine them running them along the London Underground to a patient collapsed on a station.

They will certainly transport emergency medications and equipment—such as antivenin to a remote farm after a snakebite, type AB blood to a crash site, epinephrine to a music festival. But could there be a time when dialysis machines travel the country by RPA to remote communities, to the patient?

There will likely be a time when they also transport emergency medical assistance internationally, such as organs for transplant transferred internationally by supersonic RPA; burn victims from a bombing sent skin substitutes from specialist burns hospitals across the world; vaccines and anti-viral medications delivered from the test lab, on the day, to quarantined sites.

Firefighting will also have an opportunity to gain. RPAs have already been used in firefighting, with examples of high-altitude camera use to monitor and control firefighting responses. It can go further; with fire bombers flown remotely, the risks to pilots are reduced and the aircraft can be shared internationally more readily. Perhaps swarms of 100 smaller RPAs dropping 10 liters of water each would be cheaper and more precise? If the water-bombing aircraft can be made more affordably, the fire-response services would be more empowered and the risk to firefighters could be reduced.

Policing presents some interesting concepts too. Consider applying "follow me" technology to a tennis ball-sized RPA, and setting that RPA from the bank that has just been robbed to "follow" that person. That RPA is transmitting location and video footage directly to the police, and tailing the culprit from 3–4m behind. Or tracking those on home detention arrangements.

Consider also the less comfortable idea of the police placing a gun on the RPA. For crowd control, perhaps the police could be equipped with a swarm of RPAs, each with paint guns attached, with heavy-staining paint pellets. In a mob situation, the swarm could be released to pepper the mob—the police could then identify those who were present for days afterwards simply via the dyed skin.

Emergency response teams can also look forward to the day when drones will be able to get to disaster zones faster; deliver emergency packages in bad weather, lift beams and dig snow to get to survivors in inaccessible areas and enter infected or toxic zones to direct people to safer areas. Perhaps we all look forward to that.

Video (killed the radio star)?

What will the camera-equipped RPA do to television, cinema and video BluRay? Quite possibly nothing much. There is little that an RPA can do that a helicopter or overhead boom can't do—it may just do it cheaper, and there may be more footage "at altitude" than ever before. Will it change the media we watch? If we wanted those views and perspectives, wouldn't the media production houses already utilize the altitude angle more?

There is no argument that our news is already becoming more immediate and "in the thick of it" with the equipping of drones, and this capability will increase—whether journalists are equipped with a camera drone or the public are the ones equipped; or a combination of both is used to gain coverage and quality. Perhaps over time that will increase our social tolerance of how much we want to see, and in turn movies and television will become more realistic and immersive.

On the other hand, apply the technology to the cinema itself, and load the digital "movie" onto the projection capable drone, and suddenly you have a portable cinema—your local cinema is not so tied down, not constrained to have projection rooms and perhaps can hire directly to you with "large format" movie rental.

Aerial work?

The existing industry is probably very life limited. While there is much that conventional aircraft can still do in relation to aerial work that a drone cannot, the rate of development will probably bring drones up to speed in a matter of years. Social acceptance, or not, of drones doing some work may save the aerial work industry, but I can't find an area where they won't conceivably be (or are being) overtaken—crop spraying, aerial application of insecticides and pesticides, photography, cinematography, water dropping, equipment dropping, sling loads, banners and skywriting.

On the good side of that equation, aerial work is a dangerous industry, with an uncomfortable working death rate. It will certainly be a safer and more enjoyable job to do the work via drone than risk your own life doing the same job from the seat of the plane.

The darker side of drones

Let this discussion be prefaced by the acknowledgment that agencies are already implementing anti drone tactics, such as trained falcons and netting, to enable drone seizure and disabling if needed.

In writing this, the intention is not to give people any ideas, nor to delve into intelligence roles. However, a future prediction isn't complete without looking to its dark side. Five real opportunities for misuse stand out.

First, and ugliest, is their use for terrorism. Hacking into the drone controls is actually harder than you would imagine, but purchasing your own drone and reprogramming it can be remarkably simple. This was covered more than adequately in Chapter 7.

Second, their use for illegal activities—just as a drone can deliver pizza, it can deliver drugs. Just as a drone can enter an unsafe area, a drone can enter other areas as well—and rather than drop supplies, it can lift jewellery and televisions. Our enforcement agencies will have a bigger challenge—it's one thing to identify the drone, even capture it, but tracing it back to an owner successfully can be its own challenge. Sure, some run off mobile phone apps and can be easily traced, but others can be controlled by radio alone and not matched to the controlling equipment. In fact, many can be controlled from anywhere on the planet via satellite data-links.

Third, if drones become commonplace in our lives, injuries and deaths from drone-related accidents are going to become a feature of our lives too. How great a feature that is will be influenced by our social tolerance for the incidents, particularly early on, which will drive how great the risk is that operators are allowed to take. Will we, one day, have a "drone toll" like we now have a road toll?

Fourth, take the drones into serious warfare. While they are already in use in war, imagine the likes of *Star Wars* (R2-D2 flying the aircraft for Luke) or drones fighting as in Battlestar Galactica? Or gang wars fought with drones. Truly, war would be so much more about resources and finances than ever.

Finally, on the counter side of this, micro-drones (mosquito-sized already exist, imagine them even smaller) could attach themselves to a person's clothing and be used for spying. Great if it's gathering evidence for prosecution, but not so much when it enters the board room for corporate espionage, or the hotel room for black mail or the paparazzi!

The light and fanciful side of drones

Just as the technology can be abused for damage, it can also be utilized for creative and enjoyable pastimes. Drone racing is already a developing sport. But stretch your thinking—what games would be more fun/crowd pleasing in three dimensions, or on a large scale? 3D chess at the local airshow or international chess competition? Imaginary games come alive in space (where they can't hit us personally)—World of Warcraft, World War II re-enactments, quidditch, geocaching at altitude, drone jousting, monster trucks *vs* the flying drone, drone parkour or drone hunting (where drones replace foxes and pheasant). Dedicated and specifically equipped drones won't be far off.

Industry structure?

The evolution of aviation to better incorporate unmanned aircraft should also drive changes to the structure of the aviation industry. New industries around drones are already cropping up, with specialist drone consulting, insurances and software services. One thing is certain for the UAV industry structure; there is already one growing and it will continue to grow mid-term.

As noted earlier, there are indications that the aerial work sector of aviation will be transformed to newer unmanned platforms. New uses for aviation are arising, with RPAs at weddings and micro-mapping of farms well established. So the rise of RPAs is already influencing the structure of the aviation industry. In doing so, it is pushing and testing the existing thinking that derived those structures, and will either fold into the existing structures in the way the jet engine become part of the accepted technology, or it will alter the definitions and management of the industry leading to dramatic changes in the way we view and manage the industry and how safe we expect it to be.

But what shape will the industry be next? What if RPAs become household items? The industry could well develop into just another item to be manufactured. Perhaps sold alongside your jack hammer at the hardware store, as a tool; as an extra on your car to see around corners; sold in the local department store alongside vacuum cleaners and washing machines (they already are, but in the toy section); sold with your camera as an extra fitting like a telescopic lens? They might become a house inclusion when you build new, with a docking point, fitted out and equipped to mow, mop, clean the gutters and wash the windows.

Alternately, consider what happens if larger RPAs become more accepted. How many competitors for larger RPAs can the industry maintain? While there are hundreds of operators cropping up, it remains to be seen if they will begin to consolidate, and to what extent. We may also get down to just a small number of manufacturers, who are unlikely to be the current smaller aircraft manufacturers, such as the likes of Piper, Beechcraft and Gippsland Aeronautics, as these manufacturers are heavily and expensively geared, making nimble change difficult. Some of the larger and sport manufacturers are working within the space so may come to the fore. Consolidation of operations and manufacturing will be heavily influenced by the ways in which the industry is regulated. Should a base cut-off weight be implemented below which operations don't require unique approval? Operators and manufacturers will be able to enter and flourish quite affordably, probably supporting a healthy but higher turnover industry. Heavy regulation on the other hand might force consolidation to keep costs and risks under control; too heavily regulated and an oligopoly of operations is possible. Regulation on the other hand, based on operation type, may drive consolidation into sport and hobby operations, and commercial operations. Regulation that allows for testing and experimentation has the potential to either hinder the technology growth and subsequent industry evolution, or aide it.

Rate of change of technology—how will they all keep up?

Another large influencer on the future of RPAS is the rate of change of technology—robotics, software, internet, materials and accessibility stand out. RPAs are not alone in this challenge—cars, televisions, even septic systems all benefit from the implementation of newer technologies and capabilities. All are challenged in that the technology is changing dramatically faster than the life of the items. The concept of planned obsolescence in some industries is now almost overtaken by unplanned obsolescence, where disruptive technologies have to be taken onboard nearly as fast and frequently as the old planned cycle anyway.

Think of your smart phone, which can run independently of your computer, but which requires software upgrades at unknown intervals. The phone also upgrades with improved hardware such as chip, service connection, lens quality and screen. At the moment you buy a new phone to upgrade the hardware. Imagine being able to retrofit your phone with the new Carl Zeiss lens, or mini–micro–infinitesimally-small dpi screen.

The RPA industry, with multiple kits, software, remote controls, upgrades and options, has the same challenge. For smaller, affordable RPAs, you may well upgrade like the phone. But for larger kit, the RPA industry is going to be a leading industry in coming to terms with compatibility, retrofit and upgrade management. With the airline industry well established with the skills to manage, but not the nimbleness, RPAs provide the nimbleness to implement the industry skills.

Looking into a crystal ball, there will be a time in the near future when other transport manufacturers, particularly car makers, will be seeking the advice of the aviation industry to improve their ability to "keep up safely" with the rate of change of technology.

Drone highways?

Predicting the existence of drone roads and highways is easy. Just as cars, trains and trams have their main routes to help keep traffic separation, so do aircraft (at somewhat of a pronounced altitude).

In order to function with and around us on a large scale, the drones will need to have similar traffic management strategies. What those strategies will be remains to be seen.

Could drones travel among the traffic, at or near road level? It complicates an existing system, but is not very much different from an unmanned car.

They could travel at a height above the existing roads, which would keep them out of the way of existing traffic, to keep things simple. The height of travel would be an issue—needing to be above streetlights to avoid them, but passing under bridges. Tunnels would be out of the question, so some alternative routes would have to be incorporated. Would they need traffic lights to help them at cross-sections or could traffic also be separated by height so that no cross-overs exist? Conventional aircraft have 500ft vertical separation, with North/South and East/West corridors. Maybe unmanned aircraft could have 50ft separation, and similar corridors.

Another option is to simply allow the development of new "highways" where the RPA traffic most naturally flows. Perhaps these will arise where the connectivity and visibility is at its best or where current road traffic is at its worst. These highways would be reasonably free so long as they remained under the 500ft that current aircraft utilize; but could develop up and around mega towers to reach higher levels for delivery of goods (and avoid elevators). If you are imagining the opening scenes of The Jetsons you are in the ball park.

Perhaps the use of the airspace will be separated by the use of the RPA— if it's carrying a human, it might stay at road level with cars (less distance to fall), and if it's carrying heavy loads it might travel above the emergency lane on highways only (unlikely to drop on anyone).

There are limits . . .

Great store has been put forward in the value of flying cameras, and many of the future forecasts out there espouse their use for crime detection, safety and forewarning. However, "eyes" don't provide all the protection you need—seeing a drug deal go down still allows the drugs to move and be consumed, the vision requires human intervention to prevent the deal, or it just supports prosecution. Seeing poachers approach an endangered rhino is only useful if you can protect the rhino, there's no use in watching it die, or relying on the drone to shoot the poachers only to have the drone shot down and lose vision as well as defense.

UAVs are not going to be anywhere near an overarching solution for anything. They are just a tool to help us, albeit one that is notably more flexible and user friendly than any other aircraft.

References

Finnegan, Phil "UAV Production Will Total $93 Billion" Teal Group, www.tealgroup. com/index.php/teal-group-news-media/item/press-release-uav-production-will-total-93-billion, August 19, 2015, accessed April 24, 2016.

Griffiths, Mark "Video Games and Health" US National Library of Medicine, National Institute of Health, PMC, www.ncbi.nlm.nih.gov/pmc/articles/PMC558687/, July 16, 2005, accessed April 4, 2016.

11 Keeping up with the drones's

How to get "into" (be part of) the drone movement

If you're changing the world, you're working on important things. You're excited to get up in the morning.

Larry Page (co-founder of Google)

If this book has stirred in you even a small modicum of inspiration to join us in this new and evolving world of RPAS, let this chapter both tempt and moderate you as you start.

Before progressing any further, emphasis needs to be given to safety. If you start to fly, at the very least follow ICAO's (ICAO, 2016) six basic tips for safe UAS operations, which apply wherever you are in the world:

- know the rules;
- fly in daylight and in good weather;
- always keep in sight;
- avoid areas with people;
- fly below 120m (400ft);
- stay clear of other aircraft and at least 5.5km (3.4 miles) from airport.

Review your country's specific rules

Some help finding your country's rules is given at the end of this chapter. But how to start? Do you start out with the full kit and start making a profit while you learn? Do you need to spend tens of thousands of dollars on flying lessons like pilots do? The answer to both options is that you could, but it wouldn't be smart. Here are seven reasonable, commonly successful and alternative entry approaches to the industry to consider.

Option one: fly by the seat

This option is recommended for anyone with concerns about hand/eye coordination. There are an extraordinary number of flight simulators available these days. The quality of the software and the reality of the simulations varies quite markedly, so you need to choose discerningly. Seek out the advice of existing aviation enthusiasts, who will be able to recommend (and probably let you sample) their favorite (which equates to "most realistic") flight simulators.

It doesn't matter particularly which aircraft type you fly in the simulation, the reality of the interaction is the important factor. Once loaded, practice and fly in your own time, safely and with no risk to your investment. You can explore the pre-loaded areas and for many simulators you can also download different flight areas, such as your local airport.

Be realistic with your assessment of your skills though. If you can take-off, fly out, return and land safely and consistently, you have the skills to move to learning with a real-life aircraft. Then it's time to move onto another option.

Option two: fly by the seat of your pants

This option is recommended for anyone who can still define themselves as a "child," or as someone who learns best by experience. Just buy an aircraft and head on out to try your hand at flying it. At its most basic, there is nothing to stop you diving on in. You can buy online or from most toy and hobby stores. Try to select a smaller and affordable aircraft, as the chances are it will take a bit of a beating as you learn to fly. You will want to be confident enough to be able to glue the pieces back together, or able to buy another aircraft if need be. With your first purchase, don't be tempted to spend money on buying anything with additions, or to which you can add later.

When you start flying the aircraft, make sure that you are flying around your own yard, on a still day and that other people aren't nearby. In most countries, you can probably also fly at your local park or sports ground when they aren't in use.

Keep the aircraft close to you. Don't try flying at a distance as it is just going to be further to walk to pick it up. Focus instead on building skills to control the aircraft in any direction, and when *it is facing* any direction. Master loops, barrel rolls, navigating in, over and around obstacles (not little brothers and sisters) and landings. Try a couple of different aircraft types too, so that you are confident the skills are yours and not the genius design of the engineers and programmers. Once those skills are solidly in hand, it's a smart time to upgrade your aircraft.

Option three: fly by the seat of someone else's pants

This option is recommended for anyone starting on a limited budget, or who wants to progress to drone sports. Seek out your nearest model aircraft club and join up. These clubs have been around for decades, and in most nations have qualified flying instructors, aircraft to rent or bring your own options, and tried and tested approaches to teaching you how to fly well.

This option ensures you build safe and appropriate flying habits and techniques in a safe but wider environment than option two. A strong club will both encourage and challenge you, helping you continue to build skills and capability. They will also know where and when your skills will be stepping over into regulated RPAS operations, and frequently have strong networks in the industry for mentors and advice.

Option four: learn and take-off

This option is recommended for anyone wanting to fly for a business, or who wishes to start their own business as quickly as possible. Dedicated RPAS schools are established around the world now. Check your local aviation regulatory authority to find out if they should be approved or certified in your country (Australia—yes (CASA, 2016), USA—yes if they are charging you (FAA, 2016), UK—yes if they are charging you (UK CAA, 2016)).

As RPAS schools are a comparatively new arrival into the industry, they are quite varied, and the regulation of their courses is still in evolution. From experience, the authors of this book recommend that you look for the following key items from your RPAS school:

- insurance;
- competency-based training that includes:
    - core flying (take-off, travel, turn, land);
    - landing at a distance with depth target (e.g. land on the sideways running line on a sports field or between the goal posts);
    - circling an object at a distance (e.g. fly around the goal post);
    - flying the aircraft backward;
- leads by example, sharing hard-won experience and learning from others;
- teaches hazard identification and risk mitigation; and
- teaches reporting of incidents.

One of the bonuses of attending an RPAS school is that most schools will have either approval to issue, or have the links to the regulator's forms to obtain, the unmanned pilot licenses you may need to fly your drone for income or profit.

Option five: fly with your business

This option is recommended for those who work in a business that could benefit from the use of drones. There are two reasonable approaches to taking up the advantages of drones within business. The first is to outsource the work to drones. The second is to develop a drone capability within the business.

This chapter is certainly not going to delve into the depth required for the change management decision faced by businesses here, such as financials, staffing impacts or cultural situations. Suffice to say, experience has shown that key drone-specific questions to consider before progressing include:

- Internal governance: can your business assure safe and controlled operations, is the added risk and liability within your organizational capacity?
- Intellectual property: do you have particular matters to protect?
- Frequency and novelty of the use of the drones: is development of technology needed specifically for the operation? And will the related skills stay current or lapse between uses?
- Is this work currently done with aircraft? Why or why not?

These considerations will most frequently help to sift through to a core starting point for the integration of drones into your business. From there, there are already consulting businesses specializing in drones who can assist with options and strategies for the change.

Option six: fly vicariously

This option is recommended for those who are more interested in the industry than the activity of flying.

You don't have to be a fly boy to be involved in the RPAS industry. As with aviation, there are thousands of people in the background supporting the pilots and keeping the industry flying: manufacturers, engineers, programmers, roboticists, universities, teachers, aviation mechanics, lawyers, photographers, marketers, even grounds keepers are involved. Admittedly, drones don't have so many cabin crew as passenger control for emergency evacuations isn't really needed where there are no passengers yet.

Take a look at your own career and skills or your dream career change; there are opportunities out there to join in. Connect up with the local flying club, or drone representative body, and start networking. Volunteer or take a deep dive in with a new job.

Option seven: fly financially

This option is recommended for those who see the growth opportunity but are already too busy to fly for themselves; also known as investors.

If you have a good eye for solid business opportunities and instinct for growth, investing in the industry is almost open slather at this stage. Drone-based businesses are already listing on stock exchanges, and solid existing businesses are setting heavily drone-reliant strategies in place, which you can use to influence your share purchasing if it makes sense. With a multitude of start-up businesses scrambling to take or maintain leading positions, direct investment dollars are usually welcomed. If you have strong business skills to mentor, or a smart partnering option, you will be well positioned to pick and choose your investment.

Investing is not a guaranteed path to riches, but it is a solid path to being involved in and influence the industry. And if you are wise in your choice, maybe it is also a path to wealth.

Check before you start—what are the local rules for RPAS?

- Worldwide: The ICAO keeps active links to websites with the information you need, where the country has them in place. www.icao.int/safety/RPAS/Pages/UAS-Regulation-Portal.aspx.
- USA: http://knowbeforeyoufly.org/ and www.faa.gov/uas/.
- Australia: www.casa.gov.au/aircraft/landing-page/remotely-piloted-aircraft-system.
- UK: www.caa.co.uk/Consumers/Model-aircraft-and-drones/Flying-drones/ and www.caa.co.uk/Commercial-Industry/Aircraft/Unmanned-aircraft/Unmanned-Aircraft/.
- Canada: www.tc.gc.ca/media/documents/ca-standards/Info_graphic_-_Flying_an_umanned_aircraft_-_Find_out_if_you_need_permission_from_TC.pdf.

References

Civil Aviation Safety Authority (CASA) www.casa.gov.au/aircraft/landing-page/remotely-piloted-aircraft-system, 2016, accessed June 10, 2016.

Federal Aviation Administration (FAA) www.faa.gov/uas/, 2016, accessed June 10, 2016.

International Civil Aviation Authority (ICAO) UAS Regulation Portal, www.icao.int/safety/RPAS/Pages/UAS-Regulation-Portal.aspx, 2016, accessed June 10, 2016.

UK Civil Aviation Authority (CAA) caa.co.uk/Commercial-Industry/Aircraft/Unmanned-aircraft/Unmanned-Aircraft/, 2016, accessed June 10, 2016.

Index

Locators in *italics* refer to figures and tables.